Hardware Architecture of Vending Machines

-- Integration of Systems Structure and Systems Behavior --

William S. Chao

Structure-Behavior Coalescence

Hardware Architecture	=	Hardware Structure	+	Hardware Behavior

CONTENTS

PREFACE

A hardware system is complex that it comprises multiple views such as strategy/version n, strategy/version n+1, concept, analysis, design, implementation, structure, behavior and input/output data views. Accordingly, a hardware system is defined as a set of interacting components forming an integrated whole of that hardware system's multiple views.

Since structure and behavior views are the two most prominent ones among multiple views, integrating the structure and behavior views is a method for integrating multiple views of a hardware system. In other words, structure-behavior coalescence (SBC) is a single model (model singularity) approach which results in the integration of multiple views. Therefore, it is concluded that the SBC architecture is so appropriate to model the multiple views of a hardware system.

In this book, we use the SBC architecture description language (SBC-ADL) to describe and represent the hardware architecture of the vending machine. An architecture description language is a special kind of system model used in defining the architecture of a hardware system. SBC-ADL uses six fundamental diagrams to formally grasp the essence of a hardware system and its details at the same time. These diagrams are: a) architecture hierarchy diagram, b) framework diagram, c) component channel diagram, d) component connection diagram, e) structure-behavior coalescence diagram and f) interaction flow diagram.

Hardware architecture is on the rise. By this book's introduction and elaboration of the hardware architecture of the vending machine, all readers may understand clearly how the SBC-ADL helps architects effectively perform architecting, in order to synergistically construct the fruitful hardware architecture.

ABOUT THE AUTHOR

Dr. William S. Chao is the CEO & founder of SBC Architecture International®. SBC (Structure-Behavior Coalescence) architecture is a systems architecture which demands the integration of systems structure and systems behavior of a system. SBC architecture applies to hardware architecture, software architecture, enterprise architecture, knowledge architecture and thinking architecture. The core theme of SBC architecture is: Architecture = Structure + Behavior.

William S. Chao received his bachelor degree (1976) in telecommunication engineering and master degree (1981) in information engineering, both from the National Chiao-Tung University, Taiwan. From 1976 till 1983, he worked as an engineer at Chung-Hwa Telecommunication Company, Taiwan.

William S. Chao received his master degree (1985) in information science and Ph.D. degree (1988) in information science, both from the University of Alabama at Birmingham, USA. From 1988 till 1991, he worked as a computer scientist at GE Research and Development Center, Schenectady, New York, USA.

Dr. William S. Chao has been teaching at National Sun Yat-Sen University, Taiwan since 1992 and now serves as the president of Association of Enterprise Architects, Taiwan Chapter. His research covers: systems architecture, hardware architecture, software architecture, enterprise architecture, knowledge architecture and thinking architecture.

PART I: BASIC CONCEPTS

Chapter 1: Introduction to Vending Machines

A vending machine is a machine that dispenses products such as snacks, beverages, alcohol, cigarettes and lottery tickets to customers automatically, after the customer inserts some valid coins into the machine [Segr02].

Behaviors of the *Vending Machine* consist of: a) behavior of *Getting_Customer_Payment*, b) behavior of *Returning_Customer_Payment*, c) behavior of *Delivering_Customer_Selection*, d) behavior of *Refilling_Product_Store* and e) behavior of *Refilling_Coin_Store*.

1-1 Behavior of Getting_Customer_Payment

The *Vending Machine* will accept coins or credit from the customer in payment for their purchase.

In the behavior of *Getting_Customer_Payment*, a customer shall use the *Coin_Receptacle* component to deposit a nickel, dime, or quarter coin. The *Vending Machine* will only initiate payment computation or product selection process after a valid coin is detected.

1-2 Behavior of Returning_Customer_Payment

The *Vending Machine* will return the customer's payment if he or she decides not to make a selection.

In the behavior of *Returning_Customer_Payment*, a customer shall use the *Return_Payment_Button* component to get his deposit payment returned.

1-3 Behavior of Delivering_Customer_Selection

The *Vending Machine* will accept product selection from the customer. Once the customer makes a selection, the *Vending Machine* will dispense the product to the customer.

In the behavior of *Delivering_Customer_Selection*, a customer shall use the *Product_Selection_Buttons* component to make his product selection. After that, the *Vending Machine* will dispense the selected product to the customer.

1-4 Behavior of Refilling_Product_Store

To ensure that most vending products are available for customers to purchase, the vendor will always check the product store to see if any vending product needs to be refilled. If this situation does happen, then the vendor will refill the product store.

In the behavior of *Refilling_Product_Store*, a vendor will refill some vending products into the *Product_Store* component.

1-5 Behavior of Refilling_Coin_Store

In order to ensure that there are deposited coins available for change, the vendor will always check the coin store to see if any change coin needs to be refilled. If this situation does happen, then the vendor will refill the coin store.

In the behavior of *Refilling_Coin_Store*, a vendor will refill some change coins into the *Coin_Store* component.

Chapter 2: Introduction to Hardware Architecture

A hardware system comprises multiple views such as strategy/version n, strategy/version n+1, concept, analysis, design, implementation, structure, behavior and input/output data views. A systems model is required to describe and represent all these multiple views.

The systems model describes and represents the hardware system multiple views possibly using two different approaches. The first one is the non-architectural approach and the second one is the architectural approach. The non-architectural approach respectively picks a model for each view. The architectural approach, instead of picking many heterogeneous and separated models, will use only one single multiple views coalescence (MVC) model.

In general, MVC architecture is synonymous with the hardware architecture. Since structure and behavior views are the two most prominent ones among multiple views, integrating the structure and behavior views becomes a superb approach for integrating multiple views of a hardware system. In other words, structure-behavior coalescence (SBC) leads to the coalescence of multiple views. Therefore, we conclude that SBC architecture is also synonymous with the hardware architecture.

2-1 Multiple Views of a System

In general, a hardware system is extremely complex that it consists of several evolution&motivation views such as strategy/version n and strategy/version n+1 views; it also consists of various multi-level (hierarchical) views such as concept, analysis, design and implementation views; it also consists of many systemic views such as structure, behavior and input/output data views [Kend10, Pres09, Somm06].

Figure 2-1 shows that in a hardware system all these strategy/version n, strategy/version n+1, concept, analysis, design, implementation, structure, behavior and input/output data views represent the multiple views of a hardware system.

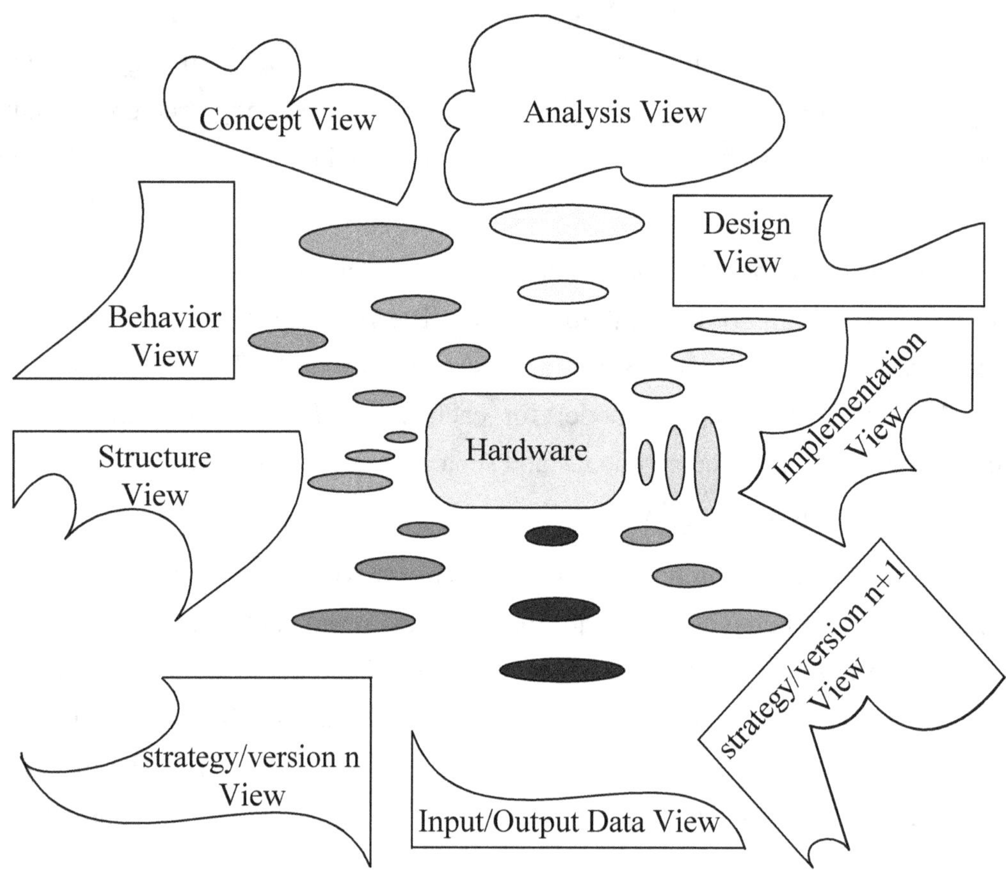

Figure 2-1 Multiple Views of a Hardware System

Among the above multiple views, the structure and behavior views are perceived as the two prominent ones. The structure view focuses on the hardware structure which is described by components and their composition while the behavior view concentrates on the hardware behavior which involves interactions (or handshakes) among the external environment's actors and components. Strategy/version n, strategy/version n+1, concept, analysis, design, implementation and input/output data views are considered to be other views as shown in Figure 2-2.

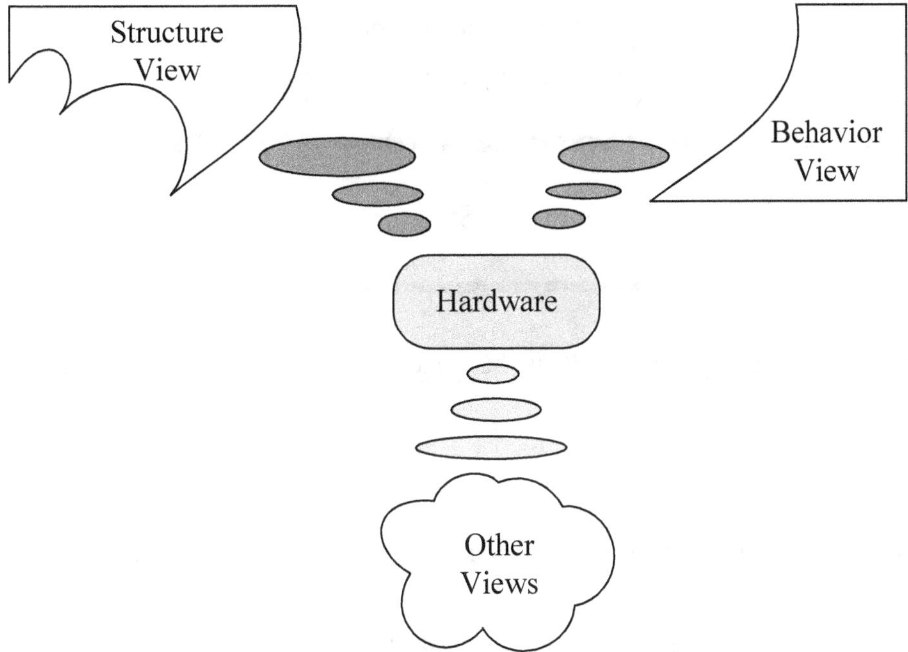

Figure 2-2 Structure, Behavior and Other Views

Accordingly, a hardware system is defined in Figure 2-3 as an integrated whole of that hardware system's multiple views, i.e., structure, behavior and other views, embodied in its assembled components, their interactions (or handshakes) with each other and the environment. Components are sometimes named as non-aggregated systems, parts, entities, objects and building blocks [Chao14a, Chao14b, Chao14c, Chec99].

A hardware system, hopefully is an integrated whole of that hardware system's multiple views, i.e., structure, behavior, and other views, embodied in its assembled components, their interrelationships with each other and the environment.

Figure 2-3 Definition of a Hardware System

Since multiple views are embodied in a hardware system's assembled components which belong to the hardware structure, they shall not exist alone. Multiple views must be loaded on the hardware structure just like a cargo is loaded on a ship as shown in Figure 2-4. There will be no multiple views if there is no hardware structure. Stand-alone multiple views are not meaningful.

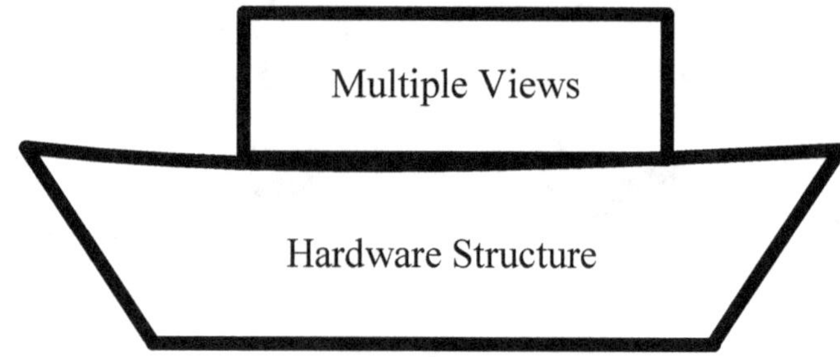

Figure 2-4 Multiple Views Must be Loaded on the Hardware Structure

2-2 Non-Architectural Approach versus Architectural Approach

A hardware system is exceptionally complex that it includes multiple views such as strategy/version n, strategy/version n+1, concept, analysis, design, implementation, structure, behavior and input/output data views.

The systems model describes and represents the hardware system multiple views possibly using two different approaches. The first one is the non-architectural approach and the second one is the architectural approach.

The non-architectural approach, also known as the model multiplicity approach [Dori95, Dori02, Dori16], respectively picks a model for each view as shown in Figure 2-5, the strategy/version n view has the strategy/version n model, the strategy/version n+1 view has the strategy/version n+1 model, the concept view has the concept model, the analysis view has the analysis model, the design view has the design model, the implementation view has the implementation model, the structure view has the structure model, the behavior view has the behavior model, and the input/output data view has the input/output data model. These multiple models are separated, always inconsistent with each other, and then become the primary cause of model multiplicity problems [Dori95, Dori02, Dori16, Pele02, Sode03].

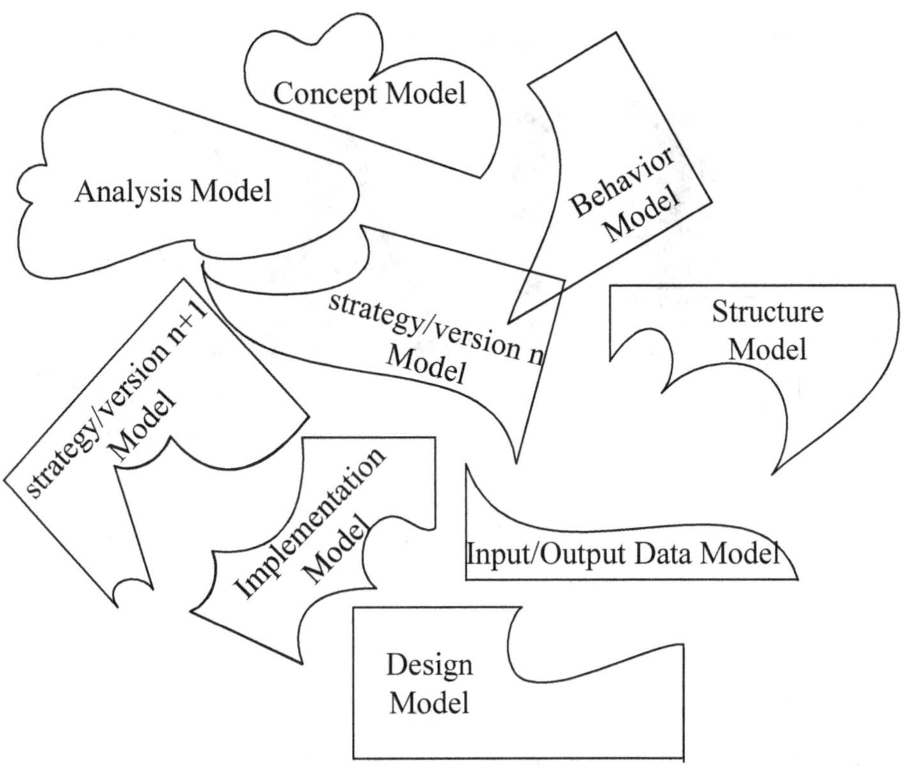

Figure 2-5 The Non-architectural Approach Picks a Model for Each View

The architectural approach, also known as the model singularity approach [Dori95, Dori02, Dori16, Pele02, Sode03], instead of picking many heterogeneous and separated models, will use only one single coalescence model as shown in Figure 2-6. The strategy/version n, strategy/version n+1, concept, analysis, design, implementation, structure, behavior and input/output data views are all integrated in this multiple views coalescence (MVC) model of systems architecture (SA) [Chao14a, Chao14b, Chao14c, Chao15a, Chao15b, Chao16, Chao17a, Chao17b, Chao17c, Chao17d, Chao17e, Chao17f].

Figure 2-6 Hardware Architecture Uses a Coalescence Model

Figure 2-5 has many models. Figure 2-6 has only one model. Comparing Figure 2-5 with Figure 2-6, we unquestionably conclude that an integrated, holistic, united, coordinated, coherent and coalescence model is more favorable than a collection of many heterogeneous and separated models.

2-3 Definition of Hardware Architecture

Involved hardware systems are extremely complex in every aspect so that each stakeholder needs a blueprint or model to capture their essential structures and behaviors. Hardware architecture is such a blueprint or model.

There are several well-know definitions of hardware architecture [Burd10, Craw15, Dam06, Maie09, O'Rou03, Roza11]. ANSI/IEEE 1471-2000 defines systems architecture as: "the fundamental organization of a system, embodied in its components, their relationships to each other and the environment, and the principles governing its design and evolution." The Open Group defines systems architecture as either "a formal description of a system, or a detailed plan of the system at component level to guide its implementation," or as "the structure of components, their interrelationships, and the principles and guidelines governing their design and evolution over time" [Rayn09, Toga08].

Concluding the above definitions, we now give hardware architecture a definition of our own as shown in Figure 2-7.

Hardware architecture is an integrated whole of a hardware system's multiple views, i.e., structure, behavior and other views, embodied in its assembled components, their interactions with each other and the environment, and the principles and guidelines governing its design and evolution.

Figure 2-7 Definition of Hardware Architecture

From the above definition, we find out that hardware architecture is an integrated whole of a system's multiple views, i.e., structure, behavior and other views, embodied in its assembled components, their interactions (or handshakes) with each other and the environment, and the principles and guidelines governing its design and evolution. That is, hardware architecture is an integrated and coalescence model of multiple views. In this coalescence model, structure, behavior and other views are all included in it as shown in Figure 2-8. We do not supply each view a respective model in this hardware architecture coalescence model.

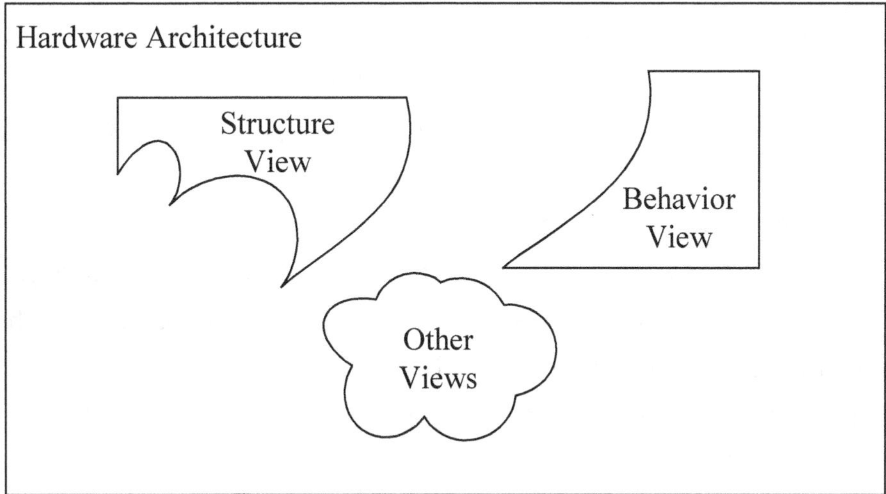

Figure 2-8 All Multiple Views are Included in This Hardware Architecture

Since multiple views are embodied in a hardware system's assembled components which belong to the structure view, they shall not exist alone. Multiple views must be loaded on the structure view just like a cargo is loaded on a ship as shown in Figure 2-9. There will be no multiple views if there is no structure view. Stand-alone multiple views are not meaningful.

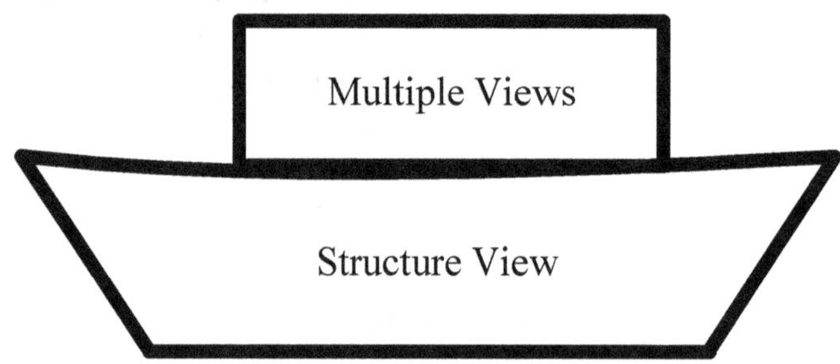

Figure 2-9 Multiple Views Must be Loaded on the Structure View

2-4 Architecture Description Language

An architecture description is a formal description and representation of a hardware system. A description of the hardware architecture has to grasp the essence of the hardware system and its details at the same time. In other words, an architecture description not only provides an overall picture that summarizes the whole hardware system, but also contains enough detail that the hardware system can be constructed and validated.

The language for architecture description is called the architecture description language (ADL) [Bass03, Clem02, Clem10, Dike01, Roza11, Shaw96, Tayl09]. An ADL is a special kind of language used in describing the architecture of a hardware system.

Since the architectural approach uses a coalescence model for all multiple views of a hardware system, the foremost duty of ADL is to make the strategy/version n, strategy/version n+1, concept, analysis, design, implementation, structure, behavior and input/output data views all integrated and coalesced within this architecture description.

2-5 Multiple Views Coalescence to Achieve the Hardware Architecture

Hardware architecture has been defined as a coalescence model of multiple views. Multiple views coalescence (MVC) uses only a single coalescence model as shown in Figure 2-10. Strategy/version n, strategy/version n+1, concept, analysis, design, implementation, structure, behavior and input/output data views are all integrated in this MVC architecture.

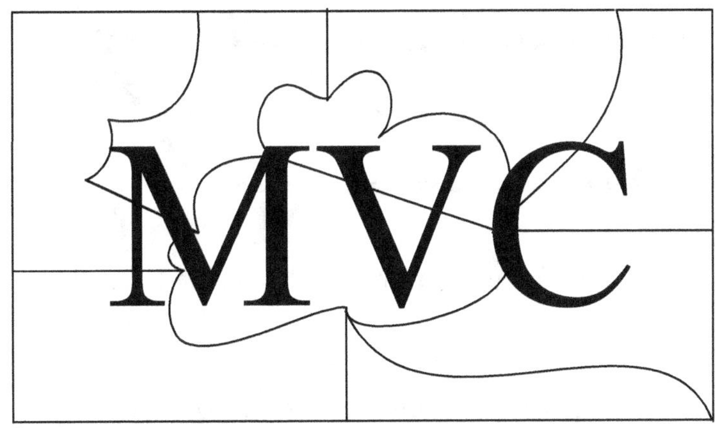

Figure 2-10 MVC Architecture

Generally, MVC architecture is synonymous with the hardware architecture. In other words, multiple views coalescence sets a path to achieve the hardware architecture as shown in Figure 2-11.

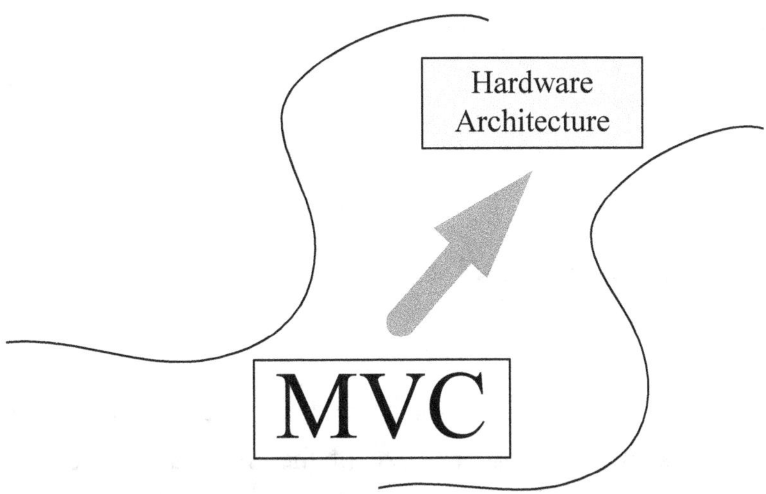

Figure 2-11 MVC to Achieve the Hardware Architecture

In the MVC architecture, multiple views must be attached to or built on the hardware structure. In other words, multiple views shall not exist alone; they must be loaded on the hardware structure just like a cargo is loaded on a ship as shown in Figure 2-12. There will be no multiple views if there is no hardware structure. Stand-alone multiple views are not meaningful.

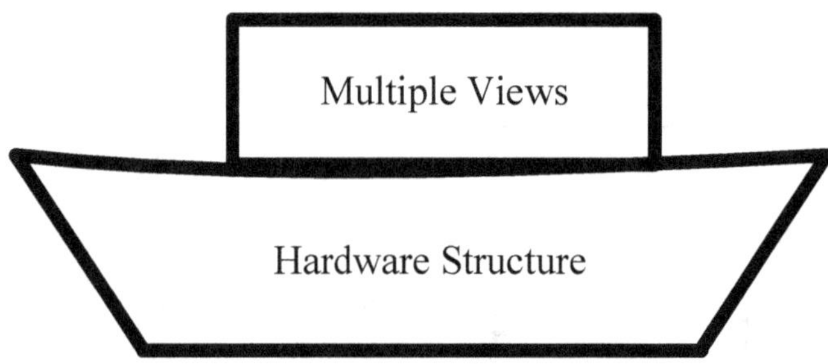

Figure 2-12 Multiple Views Must be Loaded on the Hardware Structure

2-6 Integrating the Hardware Structures and Hardware Behaviors

By integrating the hardware structure and hardware behavior, we obtain structure-behavior coalescence (SBC) within the hardware system as shown in Figure 2-13.

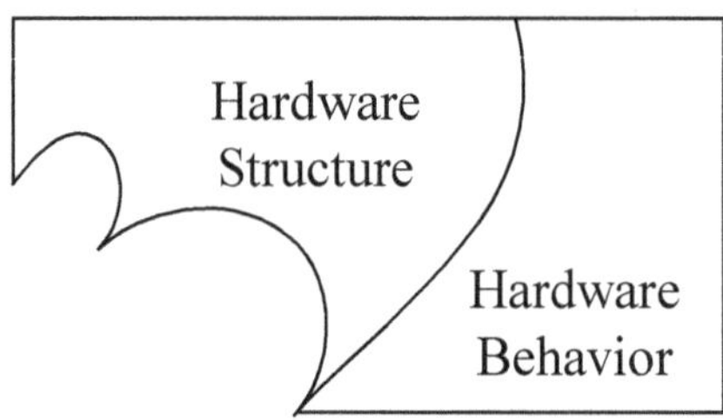

Figure 2-13 Structure-Behavior Coalescence

Structure-behavior coalescence has never been used in any systems model (SM) for hardware systems development except the SBC architecture and object-process methodology (OPM) [Dori95, Dori02, Dori16, Pele02, Sode03]. There are many advantages to use the structure-behavior coalescence approach to integrate the hardware structure and hardware behavior.

SBC architecture uses a single coalescence model as shown in Figure 2-14. Hardware structures and hardware behaviors are integrated in this SBC architecture.

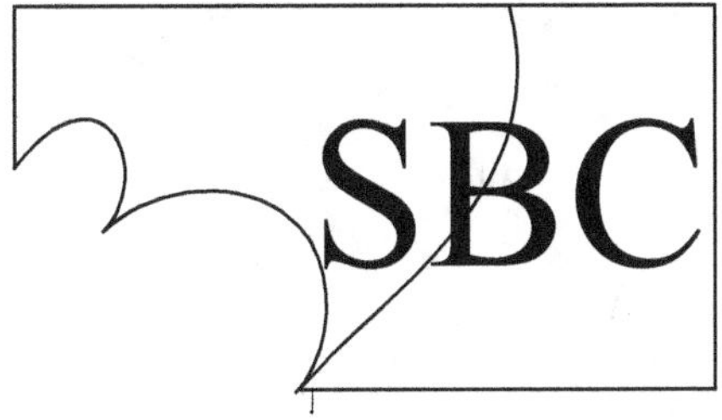

Figure 2-14 SBC Architecture

Since hardware structures and hardware behaviors are so tightly integrated, we sometimes claim that the core theme of SBC architecture is: Hardware Architecture = Hardware Structure + Hardware Behavior, as shown in Figure 2-15.

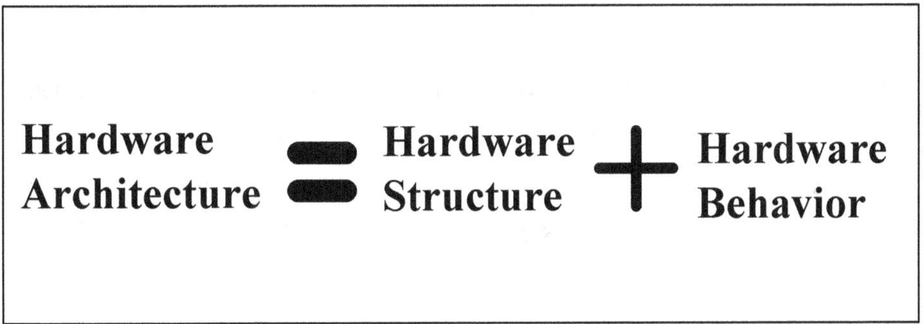

Figure 2-15 Core Theme of SBC Hardware Architecture

So far, hardware behaviors are separated from hardware structures in most cases [Pres09, Somm06]. For example, the well-known structured systems analysis and design (SSA&D) approach uses structure charts (SC) to represent the hardware structure and data flow diagrams (DFD) to represent the hardware behavior [Denn08, Kend10, Your99]. SC and DFD are two heterogeneous and separated models. They are so separated like that there is the "Atlantic Ocean" between them, as shown in Figure 2-16.

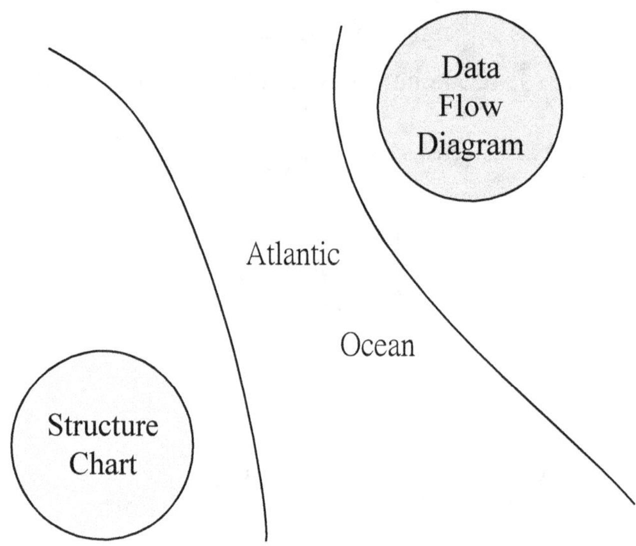

Figure 2-16 Two Heterogeneous and Separated Models

2-7 Structure-Behavior Coalescence to Facilitate Multiple Views Coalescence

Since structure and behavior views are the two most prominent ones among multiple views, integrating the structure and behavior views is clearly the best way to integrate multiple views of a hardware system. In other words, structure-behavior coalescence facilitates multiple views coalescence as shown in Figure 2-17.

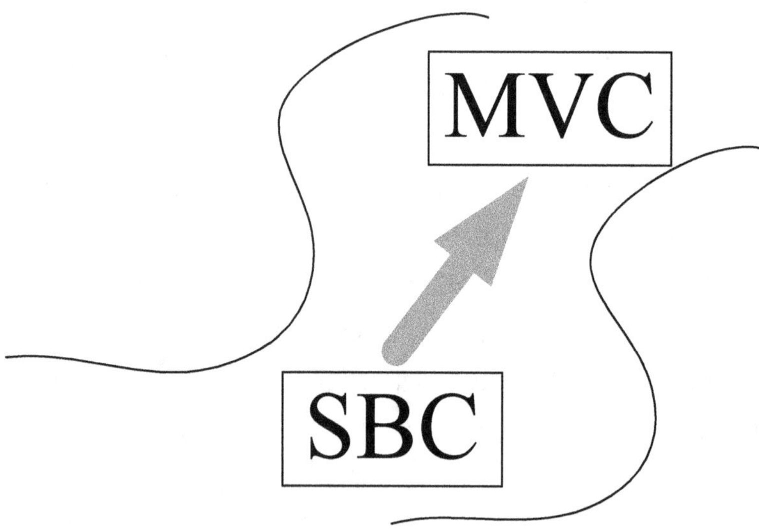

Figure 2-17 SBC Facilitates MVC

2-8 Structure-Behavior Coalescence to Achieve the Hardware Architecture

Figure 2-11 declares that multiple views coalescence sets a path to achieve the desired hardware architecture with the most efficient approach. Figure 2-17 declares that structure-behavior coalescence facilitates multiple views coalescence.

Combining the above two declarations, we conclude that structure-behavior coalescence sets a path to achieve the hardware architecture as shown in Figure 2-18. In this case, SBC architecture is also synonymous with the hardware architecture.

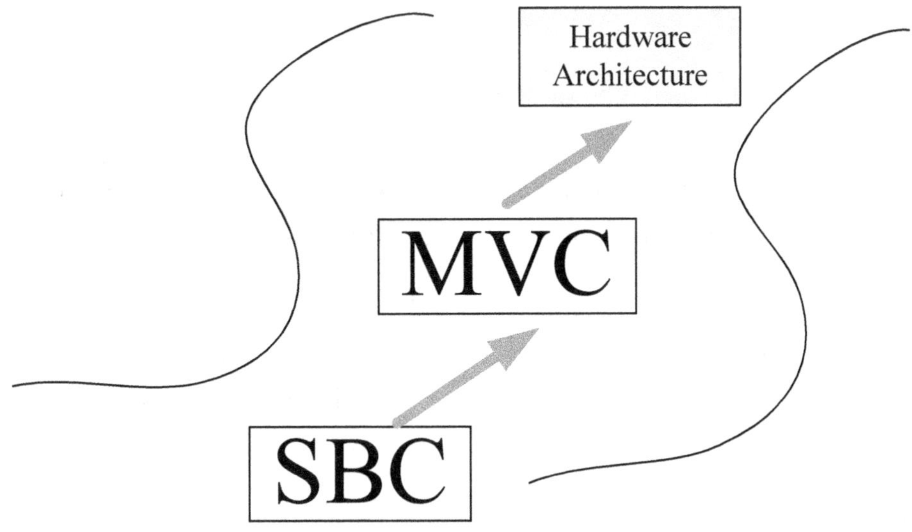

Figure 2-18 SBC to Achieve the Hardware Architecture

SBC architecture strongly demands that the structure and behavior views must be coalesced and integrated. This never happens in other architectural approaches such as Zachman Framework [O'Rou03], The Open Group Architecture Framework (TOGAF) [Rayn09, Toga08], Department of Defense Architecture Framework (DoDAF) [Dam06] and Unified Modeling Language (UML) [Rumb91]. Zachman Framework does not offer any mechanism to integrate the structure and behavior views. TOGAF, DoDAF and UML do not, either.

In the SBC architecture, the hardware behavior must be attached to or built on the hardware structure. In other words, the hardware behavior can not exist alone; it must be loaded on the hardware structure just like a cargo is loaded on a ship as shown in Figure 2-19. There will be no hardware behavior if there is no hardware structure. A stand-alone hardware behavior is not meaningful.

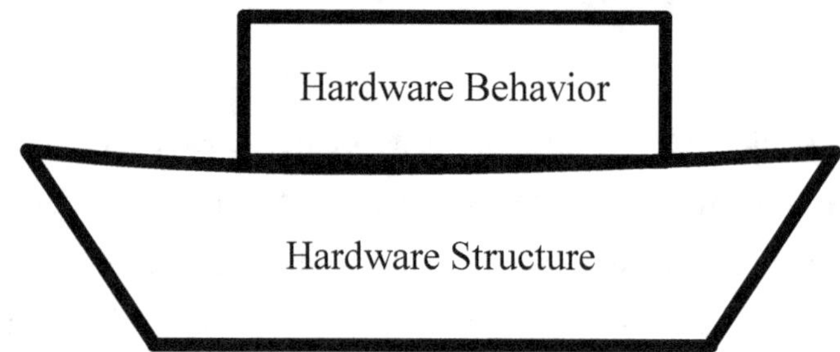

Figure 2-19 Hardware Behavior is Loaded on the Hardware Structure

2-9 Using SBC-ADL to Construct the Hardware Architecture

An architecture description language (ADL) is a special kind of language used in describing the architecture of a hardware system [Shaw96, Tayl09].

A description of the hardware architecture has to grasp the essence of a hardware system and its details at the same time. In other words, a hardware architecture description not only provides an overall picture that summarizes the hardware system, but also contains enough detail that the hardware system can be constructed and validated.

SBC-ADL uses six fundamental diagrams to describe the integration of hardware structure and hardware behavior of a system. These diagrams, as shown in Figure 2-20, are: a) architecture hierarchy diagram (AHD), b) framework diagram (FD), c) component channel diagram (CChD), d) component connection diagram (CCoD), e) structure-behavior coalescence diagram (SBCD) and f) interaction flow diagram (IFD).

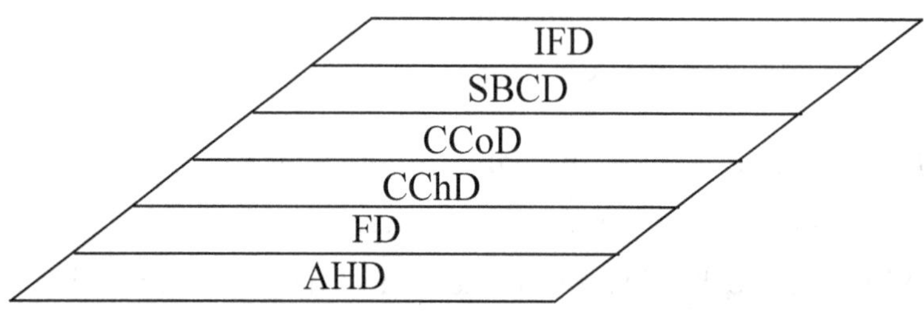

Figure 2-20 Six Fundamental Diagrams of SBC-ADL

SBC-ADL uses AHD, FD, CChD, CCoD, SBCD and IFD to depict the hardware structure and hardware behavior of a hardware system as shown in Figure 2-21.

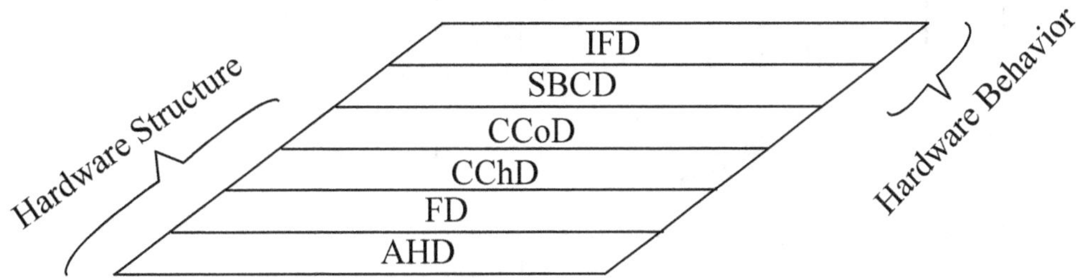

Figure 2-21 Hardware Structure and Hardware Behavior of a Hardware System

Examining the SBC-ADL approach, we find out that it depicts the hardware structure first and then depicts the hardware behavior later, not the other way around. The reason SBC-ADL does so lies in that the hardware behavior must be attached to or built on the hardware structure. With the hardware structure and attached hardware behavior, then, we can smoothly get the hardware architecture as shown in Figure 2-22.

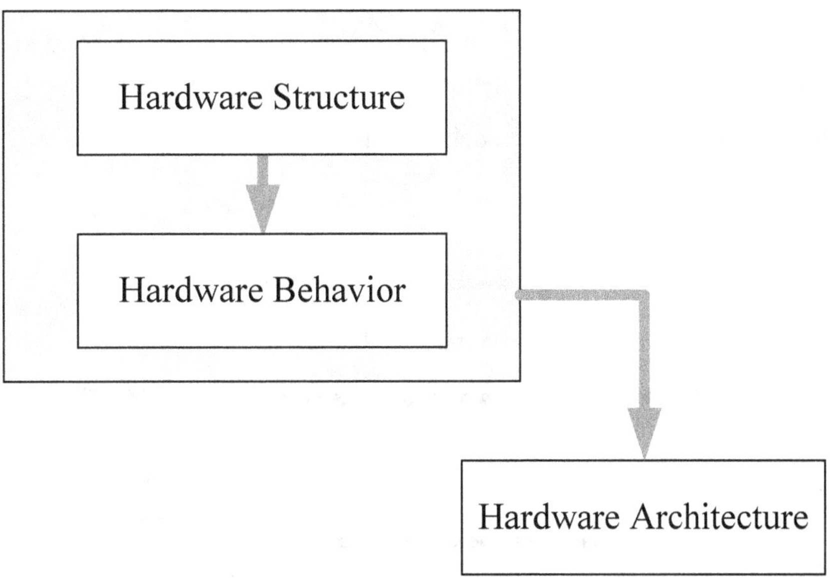

Figure 2-22 Hardware Behavior is Attached to the Hardware Structure

Let us ask the opposite question. Can the hardware structure be attached to or built on the hardware behavior? The answer is "No" as shown in Figure 2-23.

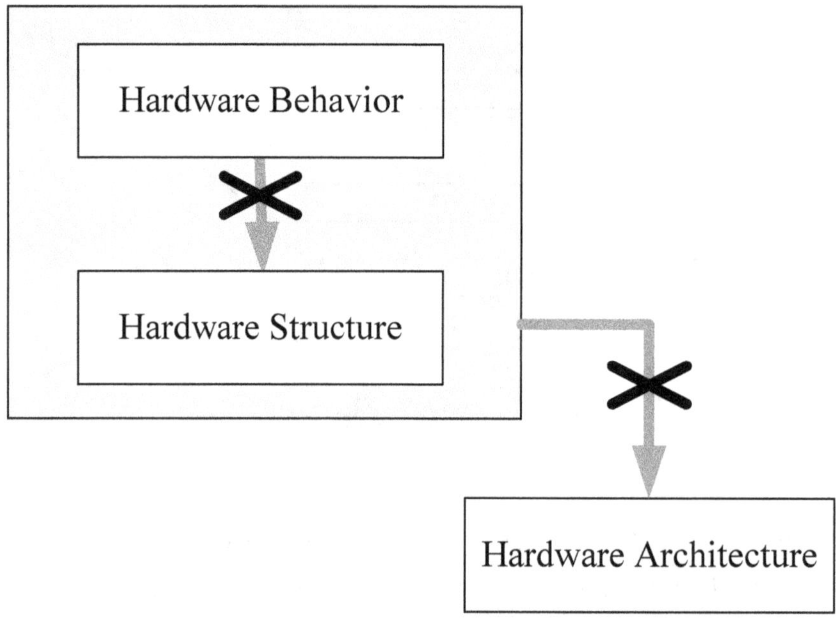

Figure 2-23 Hardware Structure is not Attached to the Hardware Behavior

In the SBC-ADL, hardware behavior must be attached to or built on the hardware structure. In other words, the hardware behavior shall not exist alone; it must be loaded on the hardware structure just like a cargo is loaded on a ship as shown in Figure 2-24. There will be no hardware behavior if there is no hardware structure. A stand-alone hardware behavior is not meaningful.

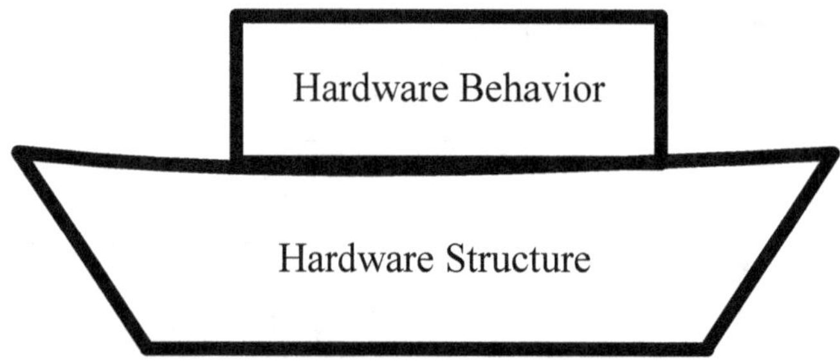

Figure 2-24 Hardware Behavior Must be Loaded on the Hardware Structure

AHD, FD, CChD and CCoD belong to hardware structure. SBCD and IFD belong to hardware behavior. Concluding the above discussion, we perceive that SBC-ADL will describe AHD, FD, CChD and CCoD first then describe SBCD and IFD later when it constructs the hardware architecture of a system..

2-10 SBC Model Singularity

Channel-Based Single-Queue SBC Process Algebra (C-S-SBC-PA) [Chao17a], Channel-Based Multi-Queue SBC Process Algebra (C-M-SBC-PA) [Chao17b], Channel-Based Infinite-Queue SBC Process Algebra (C-I-SBC-PA) [Chao17c], Operation-Based Single-Queue SBC Process Algebra (O-S-SBC-PA) [Chao17d], Operation-Based Multi-Queue SBC Process Algebra (O-M-SBC-PA) [Chao17e] and Operation-Based Infinite-Queue SBC Process Algebra (O-I-SBC-PA) [Chao17f] are the six specialized SBC process algebras. The SBC process algebra (SBC-PA) shown in Figure 2-25 is a model singularity approach.

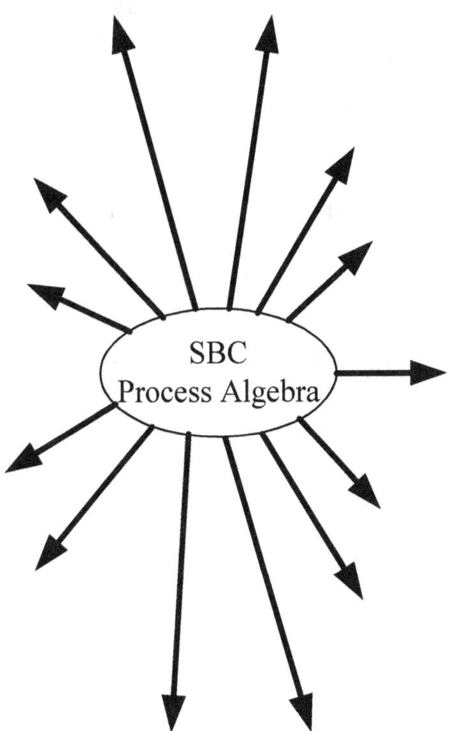

Figure 2-25 SBC-PA is a Model Singularity Approach.

The SBC architecture description language (SBC-ADL) is also a model singularity approach. With SBC mind set sitting in the kernel, the SBC-ADL single model shown in Figure 2-26 is therefore able to represent all structural views such as architecture hierarchy diagram (AHD), framework diagram (FD), component channel diagram (CChD), component connection diagram (CCoD) and behavioral views such as structure-behavior coalescence diagram (SBCD), interaction flow diagram (IFD).

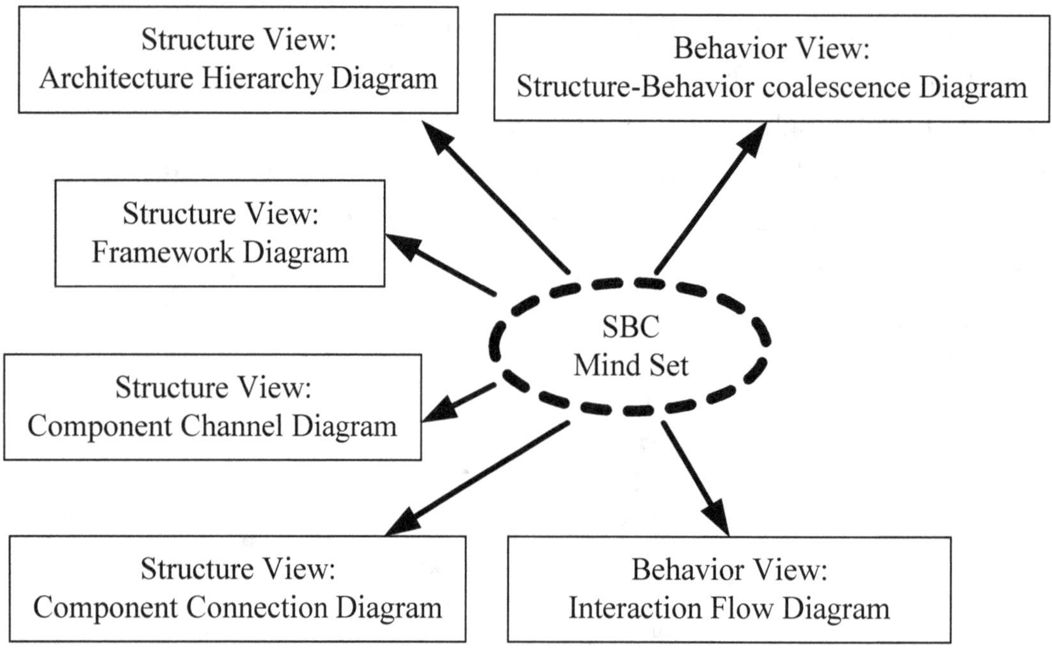

Figure 2-26 SBC-ADL is a Model Singularity Approach.

The combination of SBC process algebra (SBC-PA) and SBC architecture description language (SBC-ADL) is shown in Figure 2-27, again as a model singularity approach.

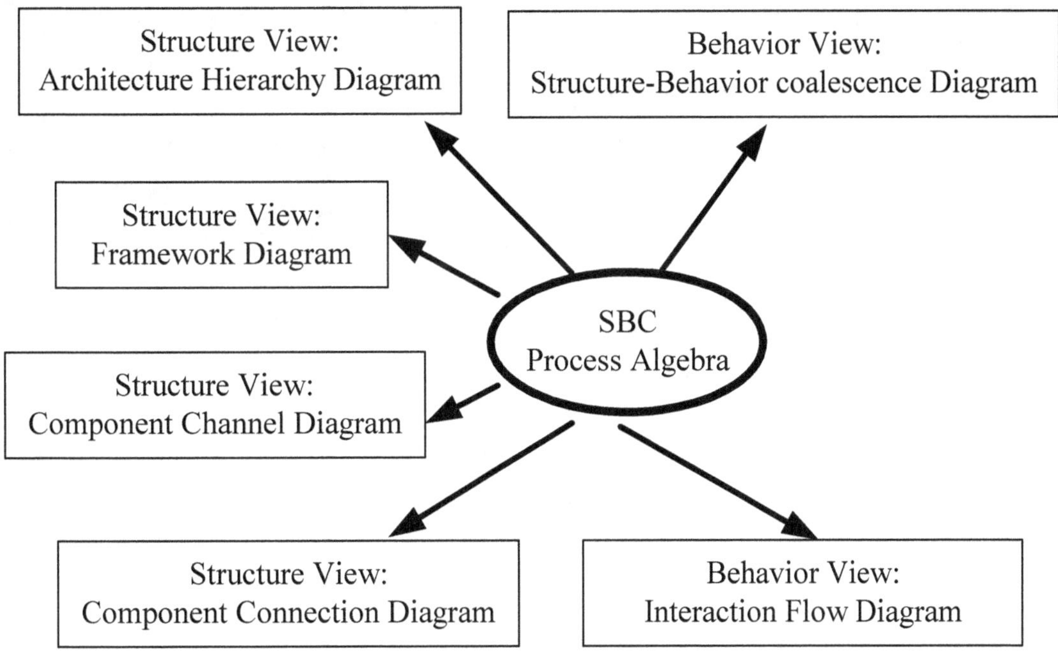

Figure 2-27 SBC Model is a Model Singularity Approach.

PART II: SBC ARCHITECTURE DESCRIPTION LANGUAGE

Chapter 3: Hardware Structure

SBC-ADL uses the architecture hierarchy diagram, framework diagram, component channel diagram and component connection diagram to depict the hardware structure of a hardware system.

3-1 Architecture Hierarchy Diagram

Hardware architects use an architecture hierarchy diagram (AHD) to define the multi-level (hierarchical) decomposition and composition of a hardware system. AHD is the first fundamental diagram to achieve structure-behavior coalescence.

3-1-1 Decomposition and Composition

The following is an example of systems decomposition and composition. The *Computer* system consists of *Monitor*, *Keyboard*, *Mouse* and *Case*, as shown in Figure 3-1. The *Monitor*, *Keyboard*, *Mouse* and *Case* are subsystems comprising the *Computer* system.

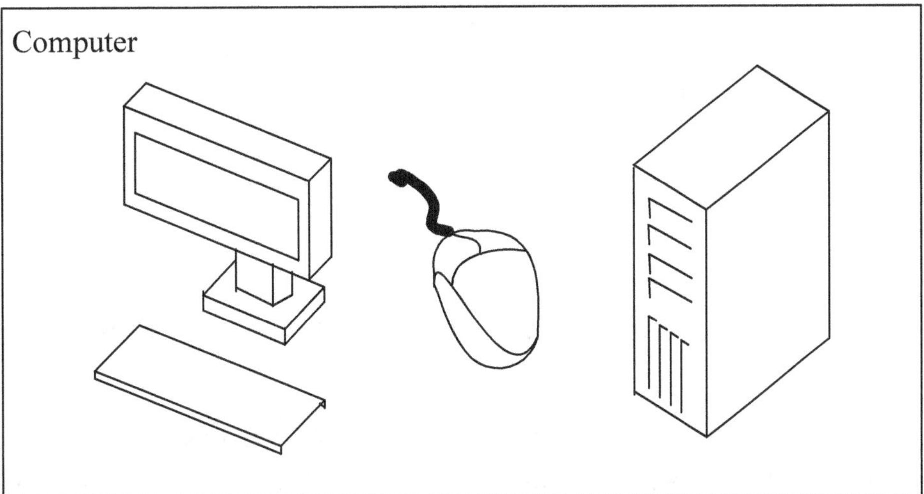

Figure 3-1 Decomposition and Composition of the *Computer* System

Another example indicates that the *Tree* system is composed of *Root* and *Stem*, as shown in Figure 3-2. In this example, we would say that the *Root* and *Stem* are subsystems, respectively, while the *Tree* system consists of its subsystems.

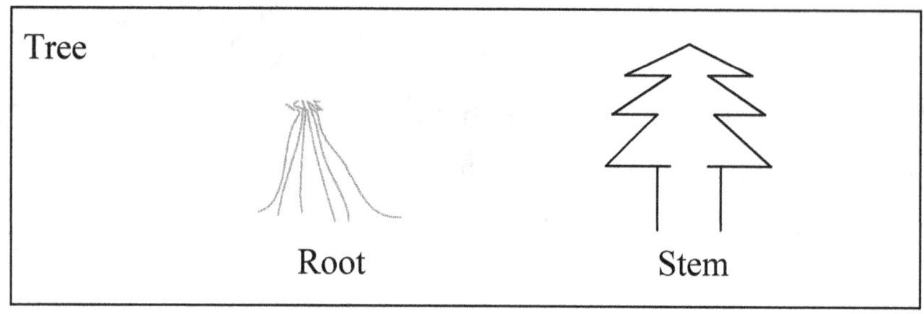

Figure 3-2 Decomposition and Composition of the *Tree* System

The last example demonstrates that the *SBC_Book* system is composed of *Chapter_1*, *Part_1* and *Part_2*, as shown in Figure 3-3. In this example, we would say that *Chapter_1*, *Part_1* and *Part_2* are subsystems, respectively while the *SBC_Book* system consists of its subsystems.

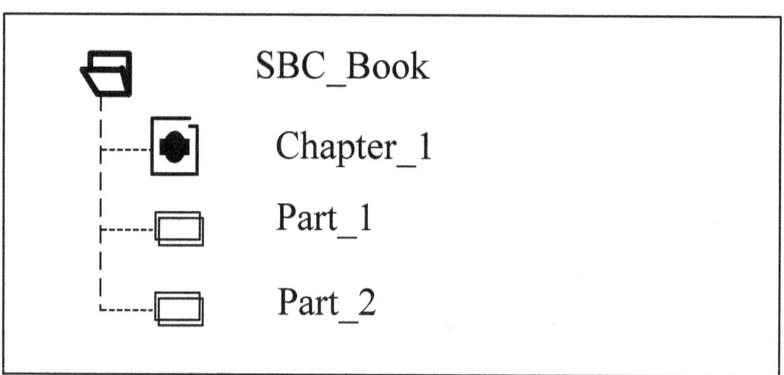

Figure 3-3 Decomposition and Composition of the *SBC_Book* System

Architecture hierarchy diagram (AHD) is used to define the decomposition and composition of a hardware system. As an example, Figure 3-4 shows an AHD of the *Computer* system. We clearly observe that the *Computer* system is composed of *Monitor*, *Keyboard*, *Mouse* and *Case*.

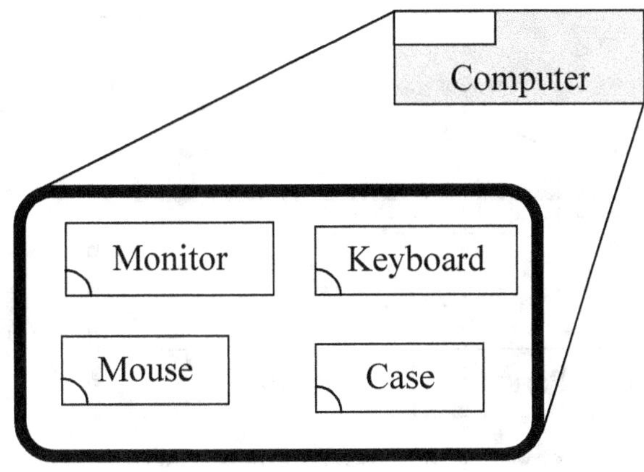

Figure 3-4 AHD of the *Computer* System

As a second example, Figure 3-5 shows an AHD of the *Tree* system. We clearly observe that the *Tree* system is composed of *Root* and *Stem*.

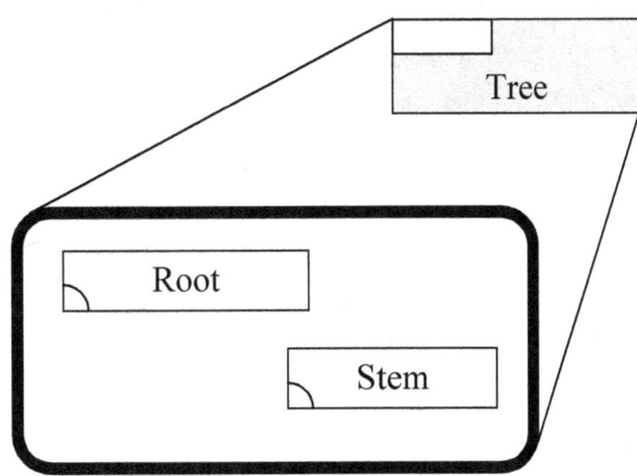

Figure 3-5 AHD of the *Tree* System

As a third example, Figure 3-6 shows an AHD of the *SBC_Book* system. We clearly observe that the *SBC_Book* is composed of *Chapter_1*, *Part_1* and *Part_2*.

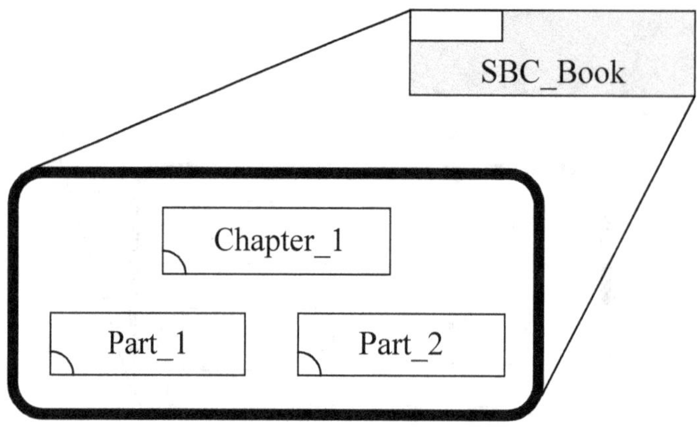

Figure 3-6 AHD of the *SBC_Book* system

3-1-2 Multi-Level Decomposition and Composition

The subsystem may also contain subsystems as we further decompose it. For example, *Case* is a subsystem of the *Computer*, and we can further decompose it into *Motherboard*, *Hard_Disk*, *Power_Supply* and *DVD_Disk*, as shown in Figure 3-7.

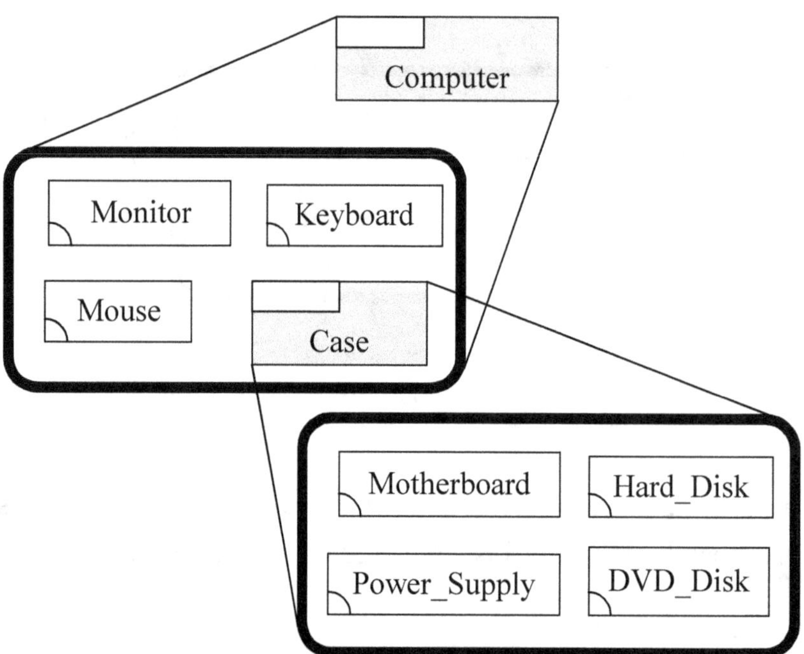

Figure 3-7 Multi-Level Decomposition/Composition of the *Computer* System

As a second example, *Stem* is a subsystem of the *Tree*, and we can further decompose it into *Trunk* and *Leaf*, as shown in Figure 3-8.

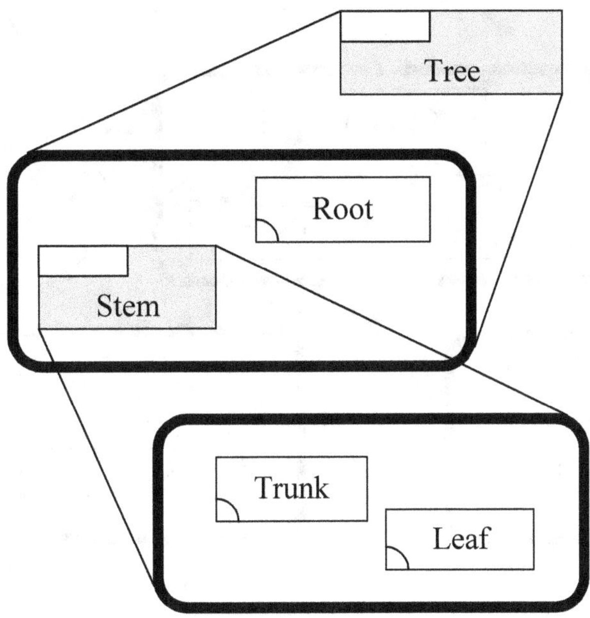

Figure 3-8 Multi-Level Decomposition/Composition of the *Tree* System

As a third example, *Part_1* is a subsystem of the *SBC_Book*, and we can further decompose it into *Chapter_2* and *Chapter_3*; *Part_2* is also a subsystem of the *SBC_Book*, and we can further decompose it into *Chapter_4* and *Chapter_5*, as shown in Figure 3-9.

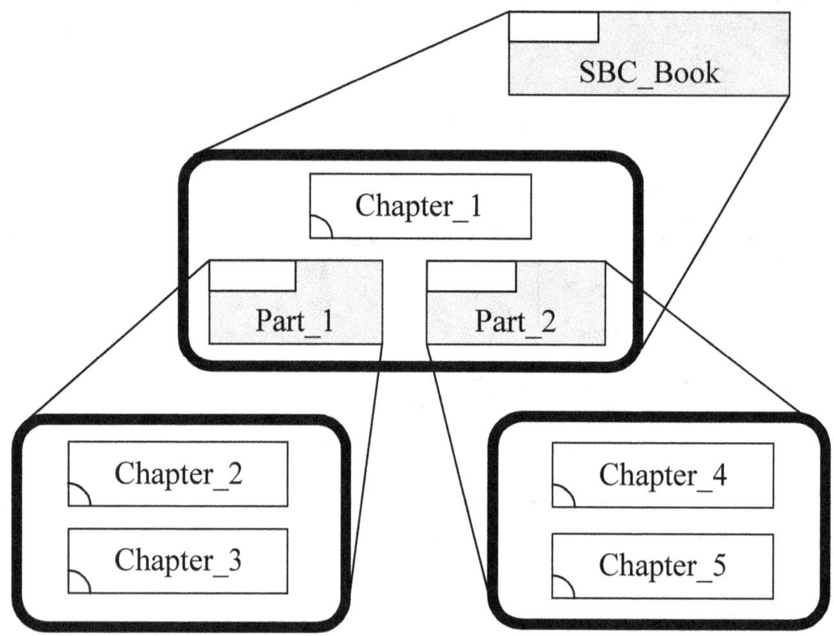

Figure 3-9 Multi-Level Decomposition/Composition of the *SBC_Book* System

Generally speaking, multi-level decomposition and composition of a hardware system is applied often in constructing its architecture. To make a complex system look simple, the mechanism of multi-level composition and decomposition should always be used.

3-1-3 Aggregated and Non-Aggregated Systems

Any subsystem (at any level) involved with multi-level decomposition and composition of a system is either aggregated or non-aggregated. The definition of aggregated and non-aggregated systems is shown in Figure 3-10.

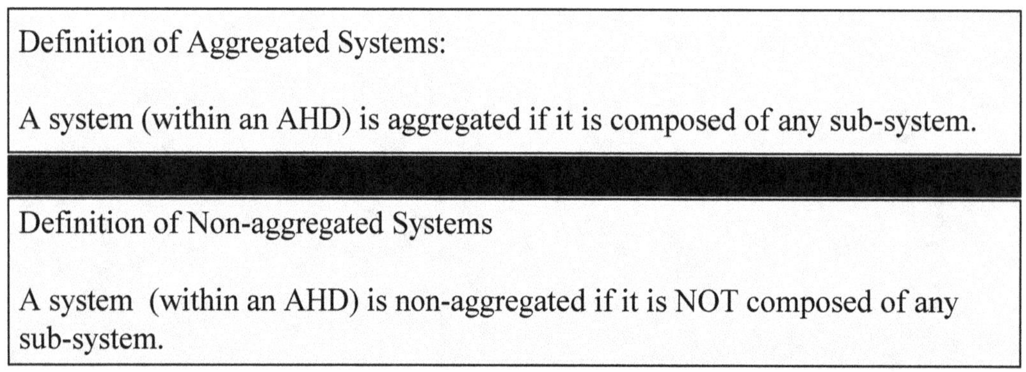

Definition of Aggregated Systems:

A system (within an AHD) is aggregated if it is composed of any sub-system.

Definition of Non-aggregated Systems

A system (within an AHD) is non-aggregated if it is NOT composed of any sub-system.

Figure 3-10 Definition of Aggregated and Non-aggregated Systems

Non-aggregated systems are sometimes referred to as components, parts, entities, objects and building blocks [Chao14a, Chao14b, Chao14c].

In the multi-level systems decomposition and composition, any system is either aggregated or non-aggregated, but not both. For example, in Figure 3-4, *Case* is a non-aggregated system, not an aggregated system. As an interesting contrast, in Figure 3-7, *Case* is an aggregated system, not a non-aggregated system.

As a second example, in Figure 3-5, *Stem* is a non-aggregated system, not an aggregated system. As an interesting contrast, in Figure 3-8, *Stem* is an aggregated system, not a non-aggregated system.

As a third example, in Figure 3-6, *Part_1 and Part_2* are non-aggregated systems, not aggregated systems. As an interesting contrast, in Figure 3-9, *Part_1* and *Part_2* are aggregated systems, not non-aggregated systems.

3-2 Framework Diagram

Framework diagram (FD) enables hardware architects to examine the multi-layer (also referred to as multi-tier) decomposition and composition of a hardware system. FD is the second fundamental diagram to achieve structure-behavior coalescence.

3-2-1 Multi-Layer Decomposition and Composition

Decomposition and composition of a system can also be represented in a multi-layer (or multi-tier) manner. We draw a framework diagram (FD) for the multi-layer decomposition and composition of a hardware system.

As an example, Figure 3-11 shows a FD of the *Computer* system. In the figure, *Technology_SubLayer_2* contains *Monitor*, *Keyboard* and *Mouse*; *Technology_SubLayer_1* contains *Motherboard, Hard_Disk, Power_Supply* and *DVD_Disk.*

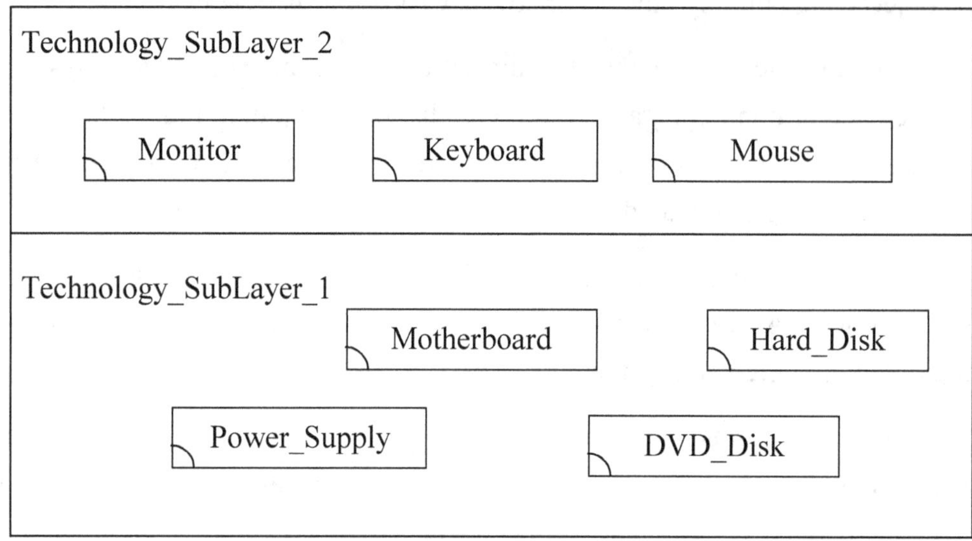

Figure 3-11 FD of the *Computer* System

As a second example, Figure 3-12 shows a FD of the *Tree* system. In the figure, *Technology_SubLayer_2* contains *Root*; *Technology_SubLayer_1* contains *Trunk* and *Leaf*.

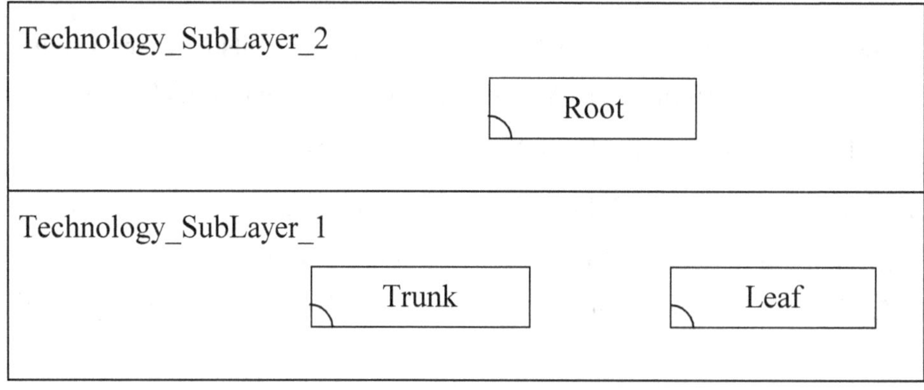

Figure 3-12 FD of the *Tree* System

As a third example, Figure 3-13 shows a FD of the *SBC_Book* system. In the figure, *Technology_SubLayer_2* contains *Chapter_1*; *Technology_SubLayer_1* contains *Chapter_2*, *Chapter_3*, *Chapter_4* and *Chapter_5*.

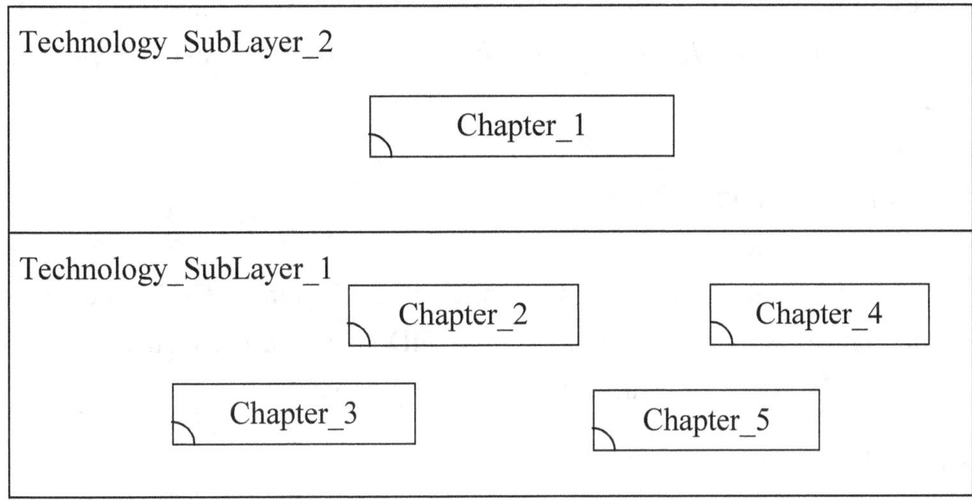

Figure 3-13 FD of the *SBC_Book* System

3-2-2 Only Non-Aggregated Systems Appearing in Framework Diagrams

Both aggregated and non-aggregated systems are displayed in the multi-level AHD decomposition and composition of a hardware system. As an interesting contrast, only non-aggregated systems shall appear in the multi-layer FD decomposition and composition of a hardware system.

For example, Figure 3-7 in the previous section shows an AHD of the *Computer* system in which both aggregated systems such as *Computer*, *Case* and non-aggregated systems such as *Monitor*, *Keyboard*, *Mouse*, *Motherboard*, *Hard_Disk*, *Power_Supply*, *DVD_Disk* are displayed. As an interesting contrast, Figure 3-11 in the previous section shows a FD of the *Computer* system in which only non-aggregated systems such as *Monitor*, *Keyboard*, *Mouse*, *Motherboard*, *Hard_Disk*, *Power_Supply* and *DVD_Disk* are displayed.

For a second example, Figure 3-8 in the previous section shows an AHD of the *Tree* system in which both aggregated systems such as *Tree*, *Stem* and non-aggregated systems such as *Root*, *Trunk*, *Leaf* are displayed. As an interesting contrast, Figure 3-12 in the previous section shows a FD of the *Tree* system in which only non-aggregated systems such as *Root*, *Trunk* and *Leaf* are displayed.

For a third example, Figure 3-9 in the previous section shows an AHD of the *SBC_Book* system in which both aggregated systems such as *SBC_Book*, *Part_1*, *Part_2* and non-aggregated systems such as *Chapter_1*, *Chapter_2*, *Chapter_3*,

Chapter_4, *Chapter_5* are displayed. As an interesting contrast, Figure 3-13 in the previous section shows a FD of the *SBC_Book* system in which only non-aggregated systems such as *Chapter_1*, *Chapter_2*, *Chapter_3*, *Chapter_4* and *Chapter_5* are displayed.

3-3 Component Channel Diagram

Hardware architects use a component channel diagram (CChD) to display all components' channels of a hardware system. CChD is the third fundamental diagram to achieve structure-behavior coalescence.

3-3-1 Channels of Components

A channel is a model for agent communication via message passing. As shown in Figure 3-14, a message may be sent over a channel, and another agent is able to receive messages sent over a channel it has a reference to.

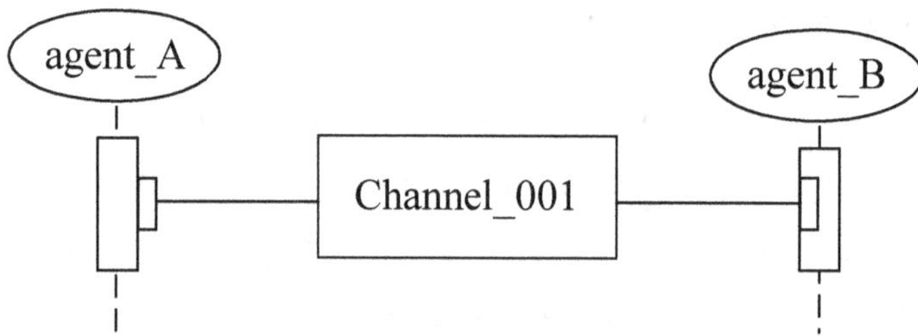

Figure 3-14 A Channel Model

A channel may contain several input parameters (e.g. i_1, i_2) and output parameters (e.g. o_1, o_2), as shown in Figure 3-15.

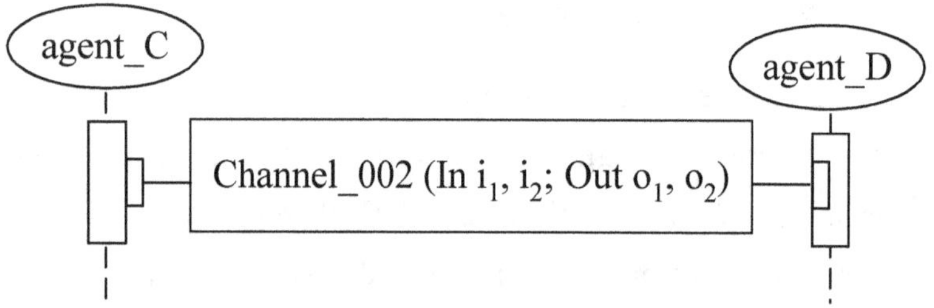

Figure 3-15 A Channel Contains Several Input/Output Parameters

A channel formula is used to completely describe a channel. A channel formula includes a) channel name, b) input parameters (e.g. $i_1, i_2, \ldots, i_m$) and c) output parameters (e.g. $o_1, o_2, \ldots, o_n$), as shown in Figure 3-16.

$$\text{Channel\_Name (In } i_1, i_2, \ldots, i_m \text{ ; Out } o_1, o_2, \ldots, o_n\text{)}$$

Figure 3-16 Channel Formula

An interaction represents an indivisible and instantaneous communication or handshake between two agents. In the channel-based approach as shown in Figure 3-17, the caller agent (either external environment's actor or component) interacts with the callee agent (component) through the channel interaction.

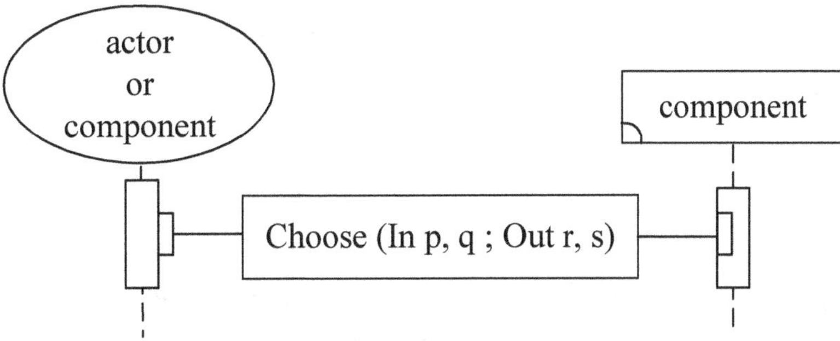

Figure 3-17 Channel-Based Value-Passing Interaction

In order to simplify the channel-based interaction diagram, we will redraw it as shown in Figure 3-18.

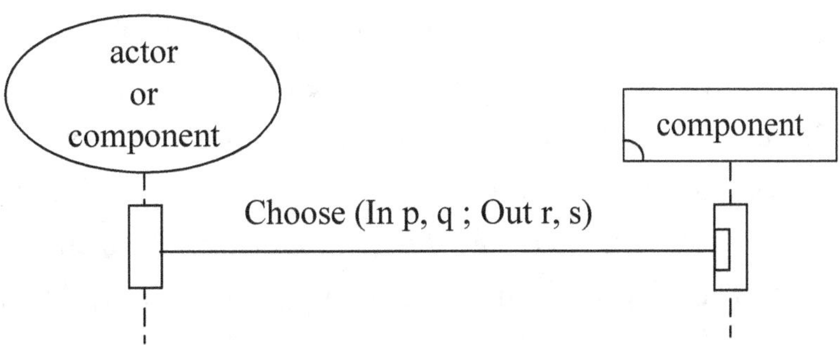

Figure 3-18 Channel-Based Interaction Diagram (I)

Or we can draw the channel-based interaction diagram as shown in Figure 3-19.

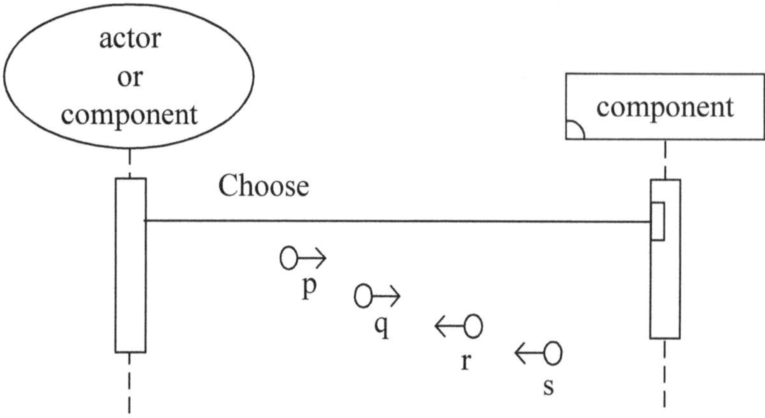

Figure 3-19 Channel-Based Interaction Diagram (II)

In general, a channel is owned by the callee agent (component). Each component in a system must possess at least one channel. A component should not exist in a system if it does not possess any channel. Figure 3-20 shows that component *SalePurchase_UI* has four channels: *SaleInputClick*, *SalePrintClick*, *PurchaseInputClick* and *PurchasePrintClick*.

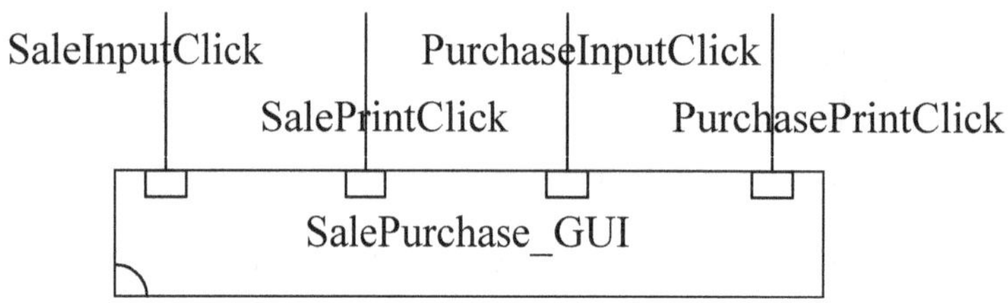

Figure 3-20 Four Channels of the *SalePurchase_GUI* Component

A channel may have several input and output parameters. The input and output parameters, gathered from all channels, represent the input data and output data views of a system [Date03, Elma10]. As shown in Figure 3-21, component *SalePrint_UI* possesses the *ShowModal* channel which has no input/output parameter; component *SalePrint_UI* also possesses the *SalePrintButtonClick* channel which has the *sDate* and *sNo* input parameters (with the arrow direction pointing to the component) and the *s_report* output parameter (with the arrow direction opposite to the component).

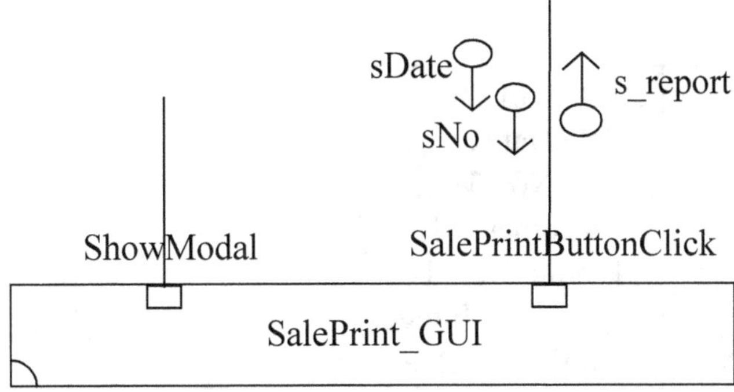

Figure 3-21 Input/Output Parameters of *SalePrintButtonClick*

Data formats of input and output parameters can be described by data type specifications. There are two sets of data types: primitive and composite [Date03, Elma10]. Figure 3-22 shows the primitive data type specification of the *sDate* and *sNo* input parameters occurring in the *SalePrintButtonClick(In sDate, sNo; Out s_report)* channel formula.

Parameter	Data Type	Instances
sDate	Text	20100517, 20100612
sNo	Text	001, 002

Figure 3-22 Primitive Data Type Specification

Figure 3-23 shows the composite data type specification of the *s_report* output parameter occurring in the *SalePrintButtonClick(In sDate, sNo; Out s_report)* channel formula.

<table>
<tr><td>Parameter</td><td>s_report</td></tr>
<tr><td>Data Type</td><td>TABLE of
Sale Date : Text
Sale No : Text
Customer : Text
ProductNo : Text
Quantity : Integer
UnitPrice : Real
Total : Real
End TABLE;</td></tr>
<tr><td>Instances</td><td>Sale Date : 20100517 Sale No : 001
Customer : Larry Fink
<table><tr><th>ProductNo</th><th>Quantity</th><th>UnitPrice</th></tr><tr><td>A12345</td><td>400</td><td>100.00</td></tr><tr><td>A00001</td><td>300</td><td>200.00</td></tr></table>Total : 100,000.00</td></tr>
</table>

Figure 3-23 Composite Data Type Specification

3-3-2 Drawing the Component Channel Diagram

For a system, CChD is used to display all components' channels. Figure 3-24 shows the *Multi-Tier Personal Data System's* CChD. In the figure, component *MTPDS_UI* has four channels: *Calculate_AgeClick_Call*, *Calculate_AgeClick_Return*, *Calculate_OverweightClick_Call*, *Calculate_OverweightClick_Return*; component *Age_Logic* has two channels: *Calculate_Age_Call*, *Calculate_Age_Return*; component *Overweight_Logic* has two channels: *Calculate_Overweight_Call*, *Calculate_Overweight_Return*; component *Personal_Database* has two channels: *Sql_DateOfBirth_Select* and *Sql_SexHeightWeight_Select*.

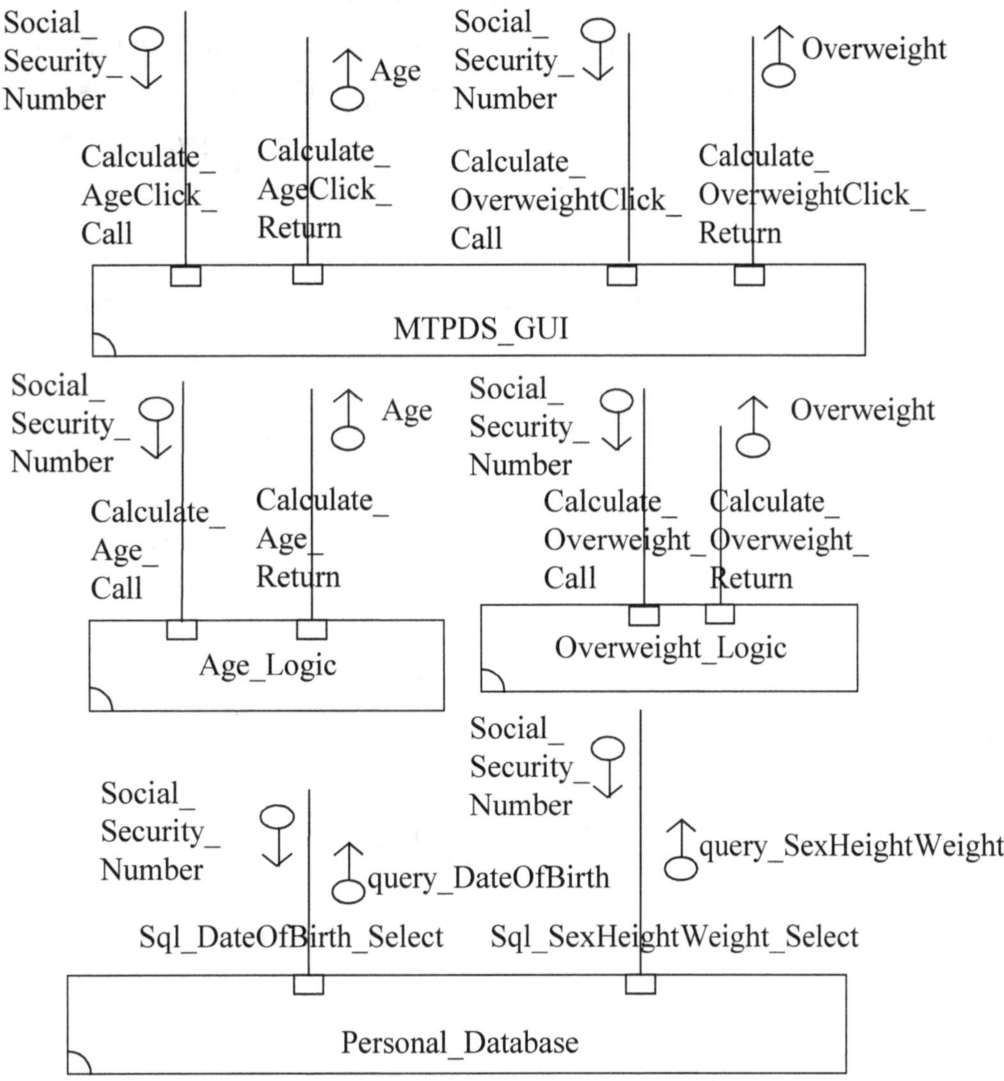

Figure 3-24 CChD of the *Multi-Tier Personal Data System*

The channel formula of *Calculate_AgeClick_Call* is *Calculate_AgeClick_Call(In Social_Security_Number)*. The channel formula of *Calculate_AgeClick_Return* is *Calculate_AgeClick_Return(Out Age)*. The channel formula of *Calculate_OverweightClick_Call* is *Calculate_OverweightClick_Call(In Social_Security_Number)*. The channel formula of *Calculate_OverweightClick_Return* is *Calculate_OverweightClick_Return(Out Overweight)*. The channel formula of *Calculate_Age_Call* is *Calculate_Age_Call(In Social_Security_Number)*. The channel formula of *Calculate_Age_Return* is

Calculate_Age_Return(Out Age). The channel formula of *Calculate_Overweight_Call* is *Calculate_Overweight_Call(In Social_Security_Number)*. The channel formula of *Calculate_Overweight_Return* is *Calculate_Overweight_Return(Out Overweight)*. The channel formula of *Sql_DateOfBirth_Select* is *Sql_DateOfBirth_Select(In Social_Security_Number; Out query_DateOfBirth)*. The channel formula of *Sql_SexHeightWeight_Select* is *Sql_SexHeightWeight_Select(In Social_Security_Number; Out query_SexHeightWeight)*.

Figure 3-25 shows the primitive data type specification of the *Social_Security_Number* input parameter and the *Age*, *Overweight* output parameters.

Parameter	Data Type	Instances
Social_Security_Number	Text	424-87-3651, 512-24-3722
Age	Integer	28, 56
Overweight	Boolean	Yes, No

Figure 3-25 Primitive Data Type Specification

Figure 3-26 shows the composite data type specification of the *query_DateOfBirth* output parameter occurring in the *Sql_DateOfBirth_Select(In Social_Security_Number; Out query_DateOfBirth)* channel formula.

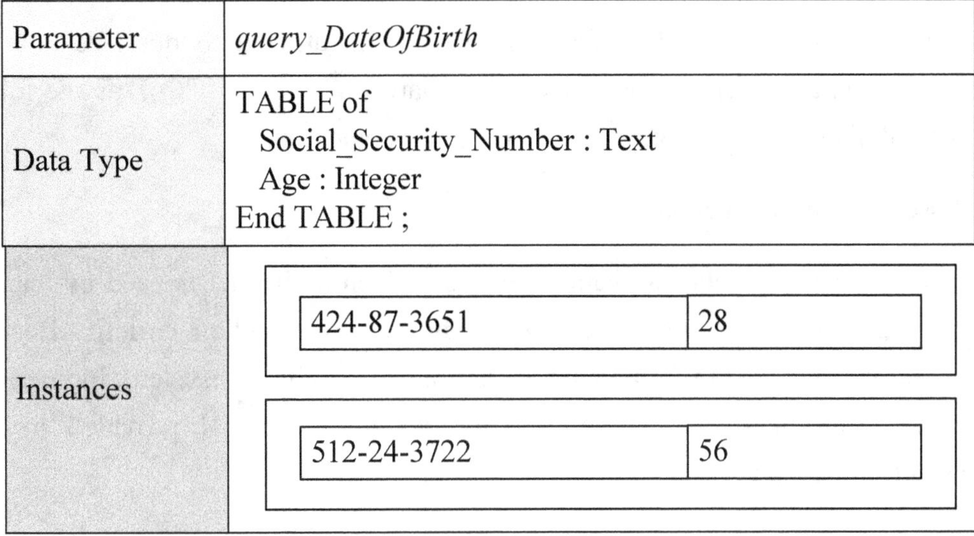

Parameter	*query_DateOfBirth*
Data Type	TABLE of Social_Security_Number : Text Age : Integer End TABLE ;
Instances	424-87-3651 | 28 512-24-3722 | 56

Figure 3-26 Composite Data Type Specification

Figure 3-27 shows the composite data type specification of the *query_SexHeightWeight* output parameter occurring in the *Sql_SexHeightWeight_Select(In Social_Security_Number; Out query_SexHeightWeight)* channel formula.

Parameter	*query_SexHeightWeight*
Data Type	TABLE of Social_Security_Number : Text Sex : Text Height : Number Weight : Number End TABLE ;
Instances	424-87-3651 | Female | 162 | 76 512-24-3722 | Male | 180 | 80

Figure 3-27 Composite Data Type Specification

3-4 Component Connection Diagram

A component connection diagram (CCoD) is utilized to describe how all components and actors are connected within a hardware system. CCoD is the fourth fundamental diagram to achieve structure-behavior coalescence.

3-4-1 Essence of a Connection

A connection implies a channel request. When a channel is used by another subsystem then a connection appears. Accordingly, a connection is defined as the linkage that is constructed when a channel is used by another subsystem. Figure 3-28 shows that Subs*ystem_A* uses the *Salary_Calculation* channel provided by the *Component_B* component.

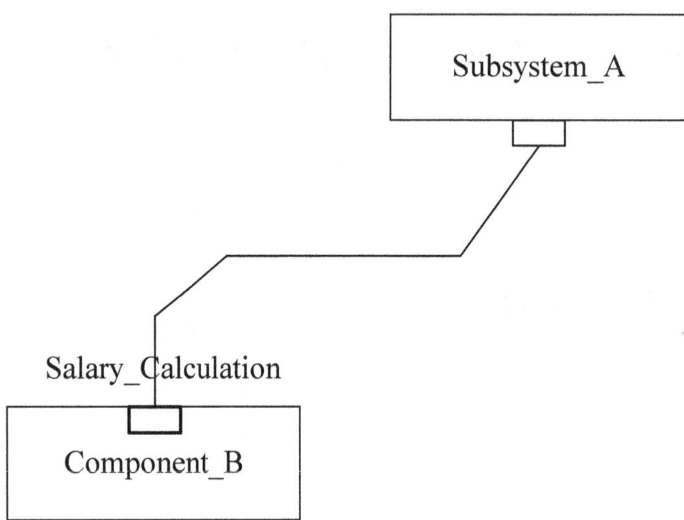

Figure 3-28 A Connection Appears When a Channel is Used

The above figure describes, sufficiently, the essence of a connection. However, we seldom use this kind of drawing. Instead, a simplified drawing of the above figure is often used as shown in Figure 3-29.

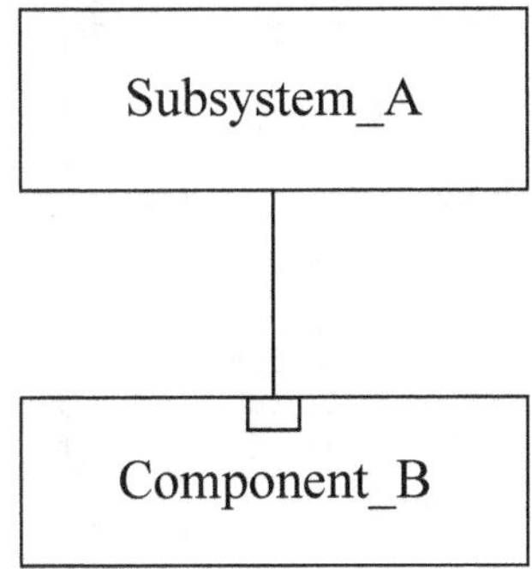

Figure 3-29 Simplified Drawing of a Connection

Since a channel is always provided by a component, there is no doubt that the *Component_B* channel provider is a component. On the contrary, the *Subsystem_A* channel user can be either a component (e.g., *Component_A*) or an actor (e.g., *Actor_A*) as shown in Figure 3-30. An actor belongs to the external environment of a system.

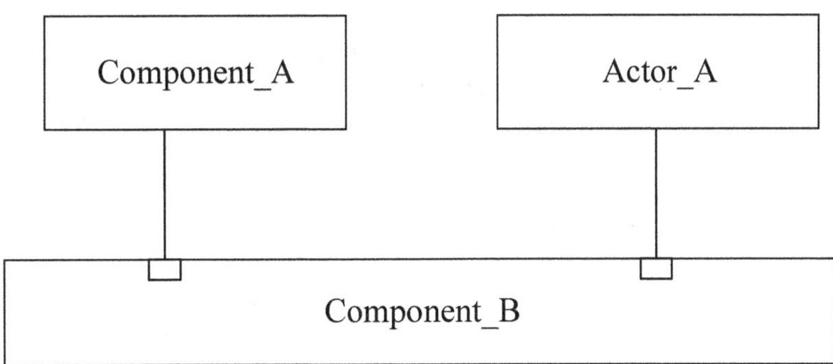

Figure 3-30 Channel User is Either a Component Or an Actor

Within a connection the subsystem (either a component or an actor) using the channel is always entitled the *Client* and the component which provides the channel is always entitled the *Server* as Figure 3-31 shows.

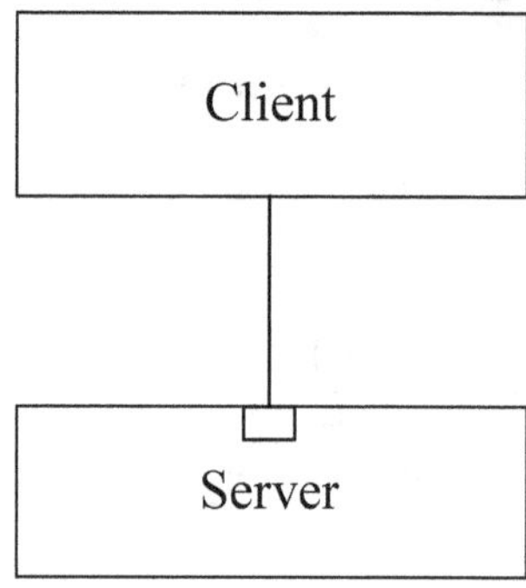

Figure 3-31 Roles of Client and Server Within a Connection

3-4-2 Drawing the Component Connection Diagram

A component connection diagram (CCoD) is utilized to describe how all components and actors (in the external environment) are connected within a hardware system. Figure 3-32 exhibits the *Multi-Tier Personal Data System's* CChD.

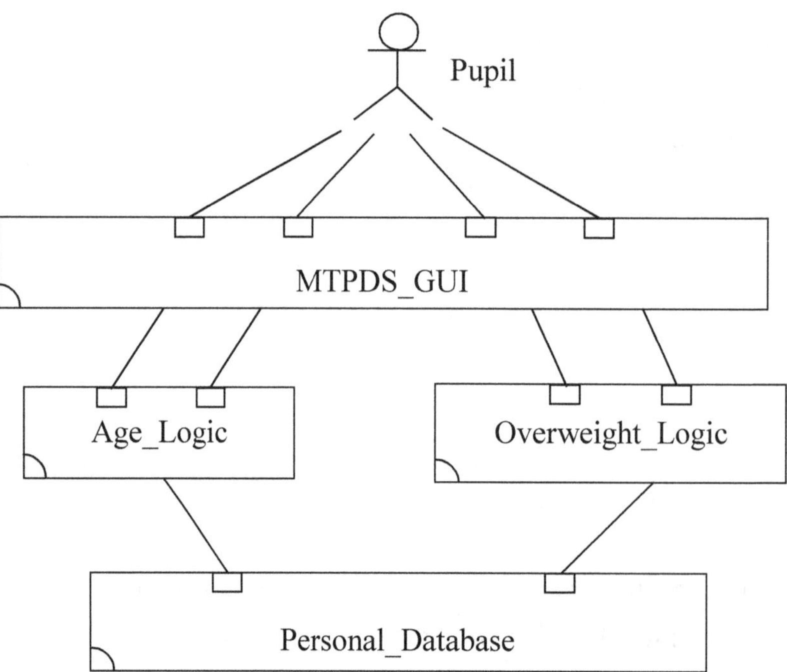

Figure 3-32 CCoD of the *Multi-Tier Personal Data System*

In Figure 3-32, actor *Pupil* has four connections with the *MTPDS_UI* component; component *MTPDS_UI* has two connections with each of the *Age_Logic* and *Overweight_Logic* components; component *Age_Logic* has a connection with the *Personal_Database* component; component *Overweight_Logic* has a connection with the *Personal_Database* component.

After finishing the CCoD, the formation pattern of the *Multi-Tier Personal Data System* will be constructed; thus the hardware structure of the *Multi-Tier Personal Data System* becomes more transparent.

Chapter 4: Hardware Behavior

SBC-ADL uses the structure-behavior coalescence diagram and interaction flow diagram to delineate the hardware behavior of a hardware system.

4-1 Structure-Behavior Coalescence Diagram

Structure-behavior coalescence diagram (SBCD) enables a hardware architect to observe the structure and behavior coexisting in a hardware system. SBCD is the fifth fundamental diagram to achieve structure-behavior coalescence.

4-1-1 Purpose of Structure-Behavior Coalescence Diagram

The major aim of the SBC-ADL approach is to achieve the integration of hardware structure and hardware behavior within a hardware system. SBCD enables a hardware architect to observe the hardware structure and hardware behavior coexisting in a hardware system. This is the purpose of utilizing SBCD when architecting the hardware architecture.

Figure 4-1 exhibits the *Multi-Tier Personal Data System*'s SBCD In this example, interactions among the *Pupil* actor and the *MTPDS_UI*, *Age_Logic*, *Overweight_Logic* and *Personal_Database* components shall draw forth the *AgeCalculation* and *OverweightCalculation* behaviors.

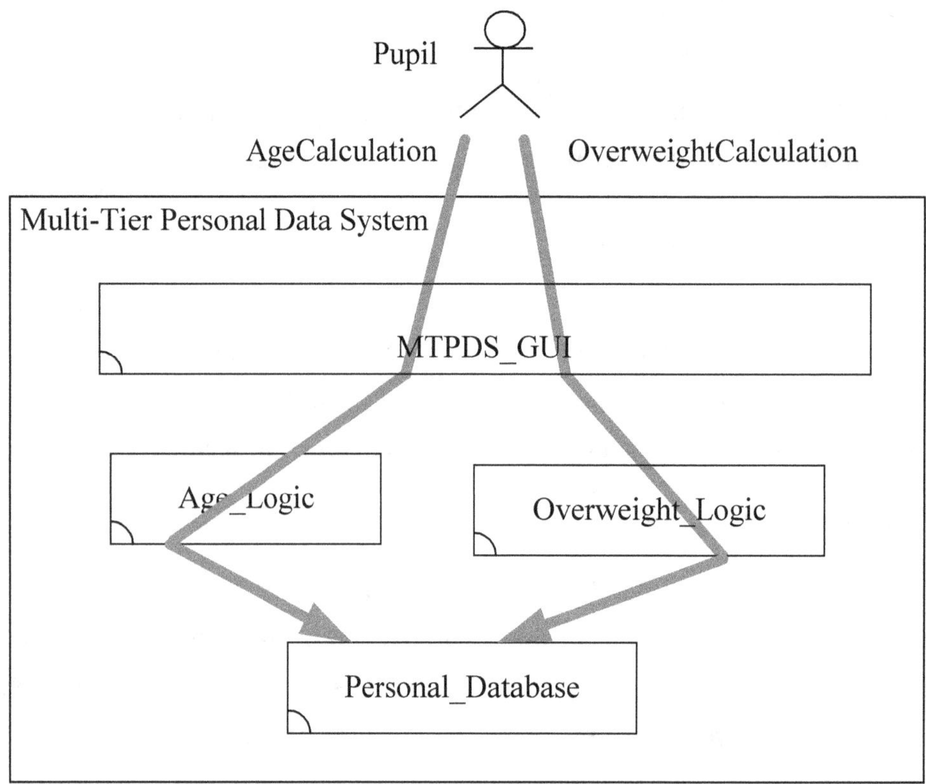

Figure 4-1 SBCD of the *Multi-Tier Personal Data System*

The overall behavior of a hardware system is the aggregation of all its individual behaviors. All individual behaviors are mutually independent of each other. They tend to be executed concurrently [Hoar85, Miln89, Miln99]. For example, the overall behavior of the *Multi-Tier Personal Data System* includes the *AgeCalculation* and *OverweightCalculation* behaviors. In other words, the *AgeCalculation* and *OverweightCalculation* behaviors are combined to produce the overall behavior of the *Multi-Tier Personal Data System*.

The major purpose of using the architectural approach, instead of separating the structure model from the behavior model, is to achieve a coalesced model. In Figure 4-1, hardware architects are able to see the hardware structure and hardware behavior coexisting in a SBCD. That is, in the *Multi-Tier Personal Data System's* SBCD, we not only see its hardware structure but also see (at the same time) its hardware behavior.

4-1-2 Drawing the Structure-Behavior Coalescence Diagram

Let us now explain the usage of SBCD by constructing a SBCD step by step. The goal of having a SBCD is enabling hardware architects to see both the structure and behavior, simultaneously. In order to achieve this goal, a SBCD is drawn by first constructing all of the components, then describing the external environment's actors, and finally describing the interactions among these components and the external environment's actors.

For example, the *Multi-Tier Personal Data System* has two behaviors: *AgeCalculation* and *OverweightCalculation.* After constructing the *Multi-Tier Personal Data System* with all its components, the external environment's actors and the *AgeCalculation* behavior, we obtain the graphical representation as shown in Figure 4-2. In this Figure, the *AgeCalculation* behavior indicates that actor *Pupil* interacts with the *MTPDS_UI* component first, then component *MTPDS_UI* interacts with the *Age_Logic* component later, then component *Age_Logic* interacts with the *Personal_Database* component finally.

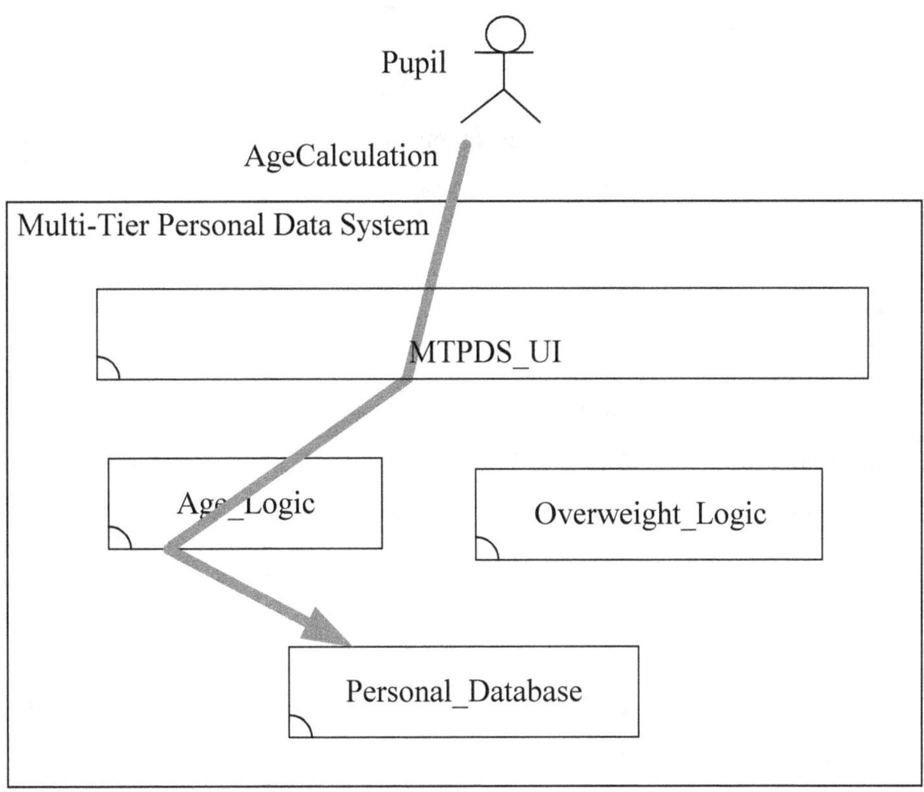

Figure 4-2 All Components, Actors, and the *AgeCalculation* Behavior

Adding the *OverweightCalculation* behavior to Figure 4-2, we then obtain the graphical representation shown in Figure 4-3. In this Figure, the *OverweightCalculation* behavior indicates that actor *Pupil* interacts with the *MTPDS_UI* component first, then component *MTPDS_UI* interacts with the *Overweight_Logic* component later, then component *Overweight_Logic* interacts with the *Personal_Database* component finally.

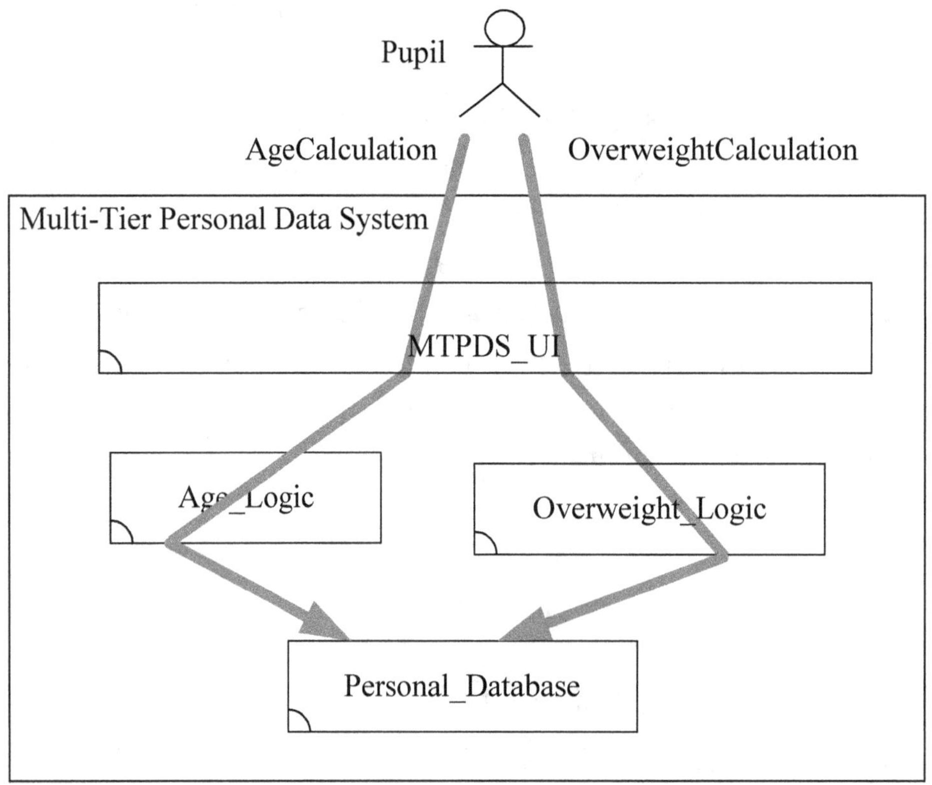

Figure 4-3 Adding the *OverweightCalculation* Behavior to Figure 4-2

After finishing Figure 4-3, we actually have accomplished all the works needed to draw an entire SBCD of the *Multi-Tier Personal Data System*. As a matter of fact, Figure 4-3 shows exactly the *Multi-Tier Personal Data System*'s SBCD.

4-2 Interaction Flow Diagram

An interaction flow diagram (IFD) is utilized to describe each individual behavior of the overall behavior of a hardware system. IFD is the sixth fundamental diagram to achieve structure-behavior coalescence.

4-2-1 Individual Hardware Behavior Represented by Interaction Flow Diagram

The overall behavior of a hardware system consists of many individual behaviors. Each individual behavior represents an execution path. An IFD is utilized to represent such an individual behavior.

Figure 4-4 demonstrates that the *Multi-Tier Personal Data System* has two behaviors; thus, it has two IFDs.

Enterprise	IFD
Multi-Tier Personal Data System	AgeCalculation
	OverweightCalculation

Figure 4-4 *Multi-Tier Personal Data System* has Two IFDs

4-2-2 Drawing the Interaction Flow Diagram

Let us now explain the usage of interaction flow diagram (IFD) by drawing an IFD step by step. Figure 4-5 demonstrates an IFD of the *SaleInput* behavior. The X-axis direction is from the left side to right side and the Y-axis direction is from the above to the below. Inside an IFD, there are four elements: a) external environment's actor, b) components, c) interactions and d) input/output parameters. Participants of the interaction, such as the external environment's actor and each component, are laid aside along the X-axis direction on the top of the diagram. The external environment's actor which initiates the sequential interactions is always placed on the most left side of the X-axis. Then, interactions among the external environment's actor and components successively in turn decorate along the Y-axis direction. The first interaction is placed on the top of the Y-axis position. The last interaction is placed on the bottom of the Y-axis position. Each interaction may carry several input and/or output parameters.

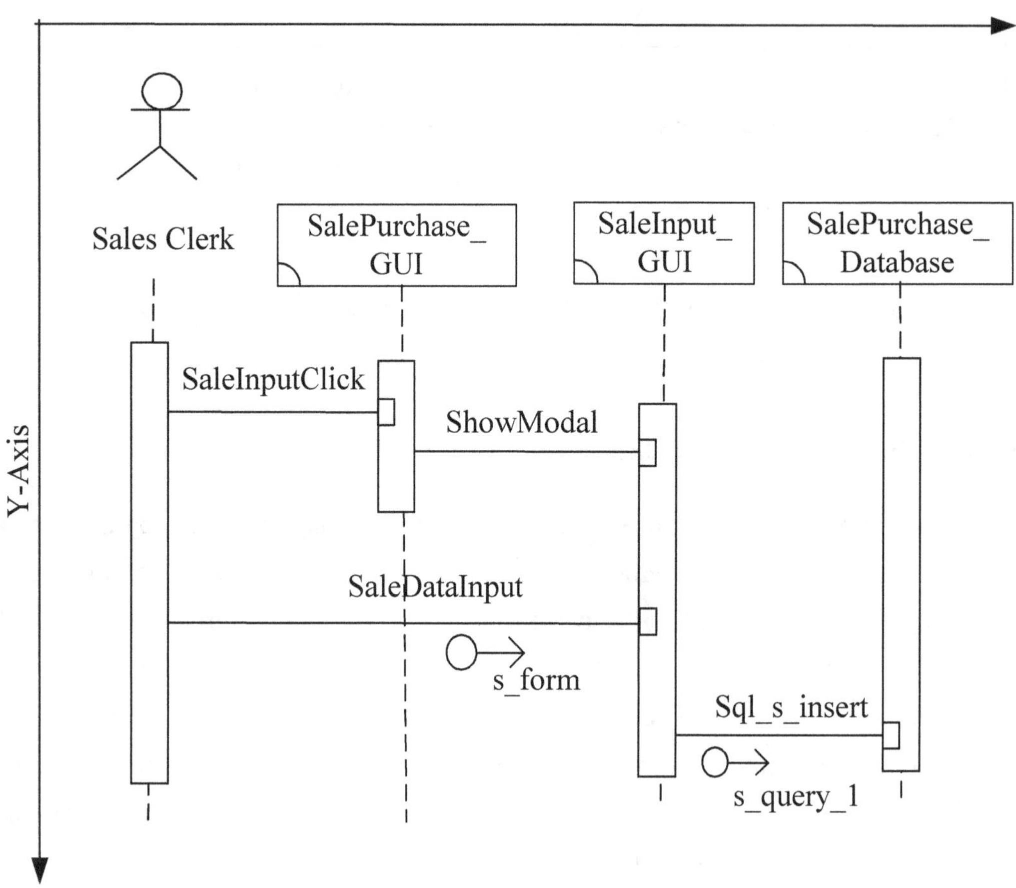

Figure 4-5 IFD of the *SaleInput* Behavior

In Figure 4-5, *Sales Clerk* is an external environment's actor. *SalePurchase_UI*, *SaleInput_UI* and *SalePurchase_Database* are components. *SaleInputClick* is a channel which is provided by the *SalePurchase_UI* component. *ShowModal* is a channel which is provided by the *SaleInput_UI* component. *SaleDataInput* is a channel, carrying the *s_form* input parameter, which is also provided by the *SaleInput_UI* component. *Sql_s_insert* is a channel, carrying the *s_query_1* input parameter, which is provided by the *SalePurchase_Database* component.

The execution path of Figure 4-5 is as follows. First, actor *Sales Clerk* interacts with the *SalePurchase_UI* component through the *SaleInputClick* channel interaction. Next, component *SalePurchase_UI* interacts with the *SaleInput_UI* component through the *ShowModal* channel interaction. Continuingly, actor *Sales Clerk* interacts with the *SaleInput_UI* component through the *SaleDataInput* channel interaction, carrying the *s_form* input parameter. Finally, component *SaleInput_UI* interacts with the *SalePurchase_Database* component through the *Sql_s_insert* channel interaction, carrying the *s_query_1* input parameter.

An interaction flow diagram may contain a conditional expression. Figure 4-6 shows such an example which has the following execution path. First, external environment's actor *Employee* interacts with the *Computer* component through the *Open_Call* channel interaction, carrying the *Task_No* input parameter. Next, if the *var_1 < 4 & var_2 > 7* condition is true then component *Computer* shall interact with the *Skype* component through the *Ch_1_Call* channel interaction and component *Skype* shall interact with the *Earphone* component through the *Ch_4_Call* channel interaction, carrying the *Skype_Earphone* output parameter; else if the *var_3 = 99* condition is true then component *Computer* shall interact with the *Skype* component through the *Ch_2_Call* channel interaction and component *Skype* shall interact with the *Speaker* component through the *Ch_5* channel interaction, carrying the *Skype_Speaker* output parameter; else component *Computer* shall interact with the *Youtube* component through the *Ch_3_Call* channel interaction and component *Youtube* shall interact with the *Speaker* component through the *Ch_6* channel interaction, carrying the *Youtube_Speaker* output parameter. Continuingly, if the *var_1 < 4 & var_2 > 7* condition is true then component *Computer* shall interact with the *Skype* component through the *Ch_1_Return* channel interaction, carrying the *Status_1* output parameter; else if the *var_3 = 99* condition is true then component *Computer* shall interact with the *Skype* component through the *Ch_2_Return* channel interaction, carrying the *Status_2* output parameter; else component *Computer* shall interact with the *Youtube* component through the *Ch_3_Return* channel interaction, carrying the *Status_3* output parameter. Finally, external environment's actor *Employee* interacts with the *Computer* component through the *Open_Return* channel interaction, carrying the *Status* output parameter.

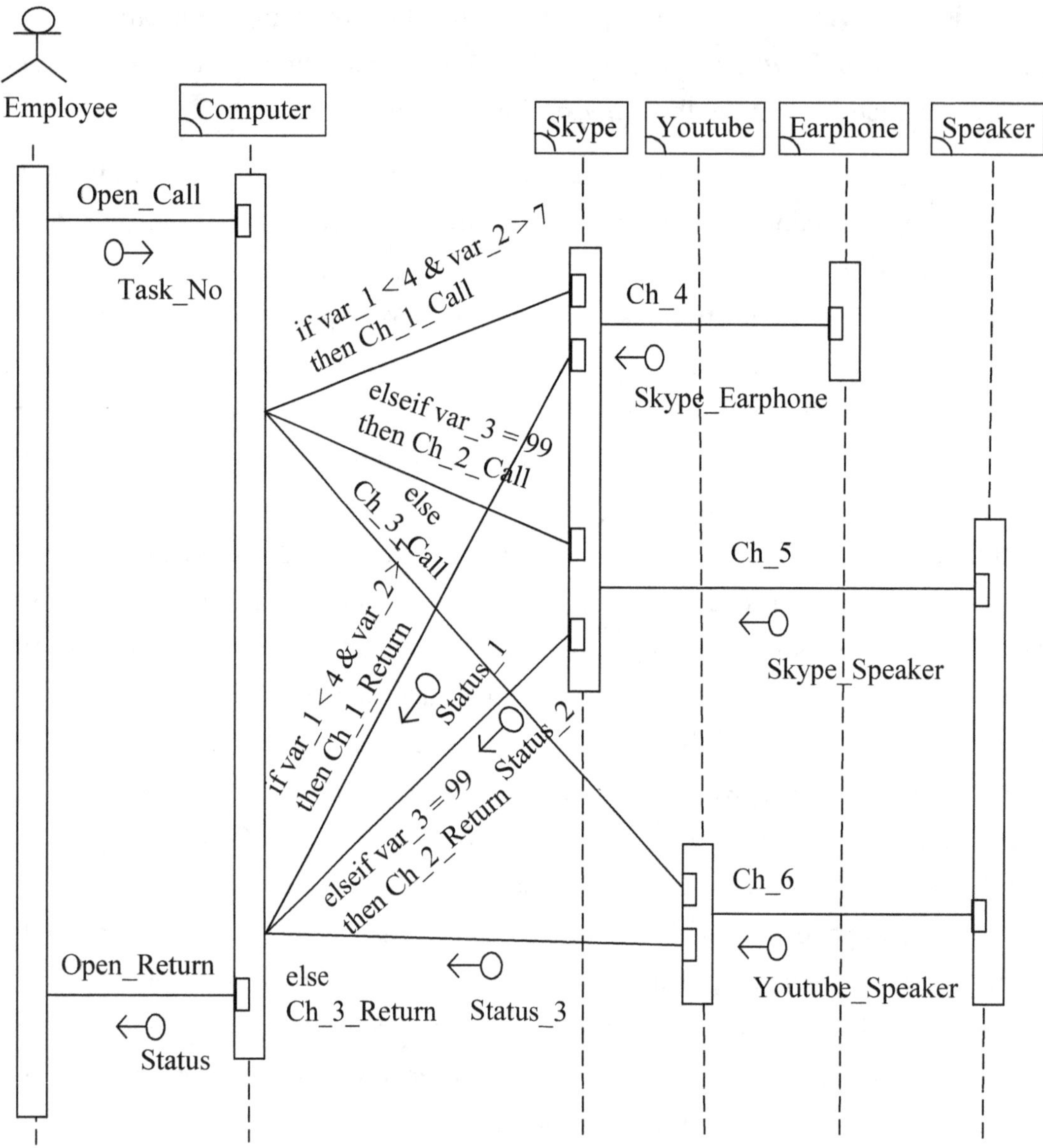

Figure 4-6 Conditional Expression

Several Boolean conditions are shown in Figure 4-6. They are "$var\_1 < 4$ & $var\_2 > 7$" and "$var\_3 = 99$". Variables, such as *var_1*, *var_2* and *var_3*, appearing in the Boolean condition can be local or global variables [Prat00, Seth96].

PART III: HARDWARE ARCHITECTURE OF VENDING MACHINES

Chapter 5: AHD of the Vending Machine

AHD is the architecture hierarchy diagram we obtain after the architecture construction is finished. Figure 5-1 shows an AHD of the *Vending Machine*. In the figure, *Vending Machine* is composed of *Interface_Layer*, *Control_Layer* and *Storage_Layer*; *Interface_Layer* is composed of *Coin_Receptacle*, *Return_Payment_Button*, *Coin_Dispenser*, *Product_Selection_Buttons* and *Product_Dispenser*; *Control_Layer* is composed of *Product_Vending_Controller*; *Storage_Layer* is composed of *Coin_Store* and *Product_Store*.

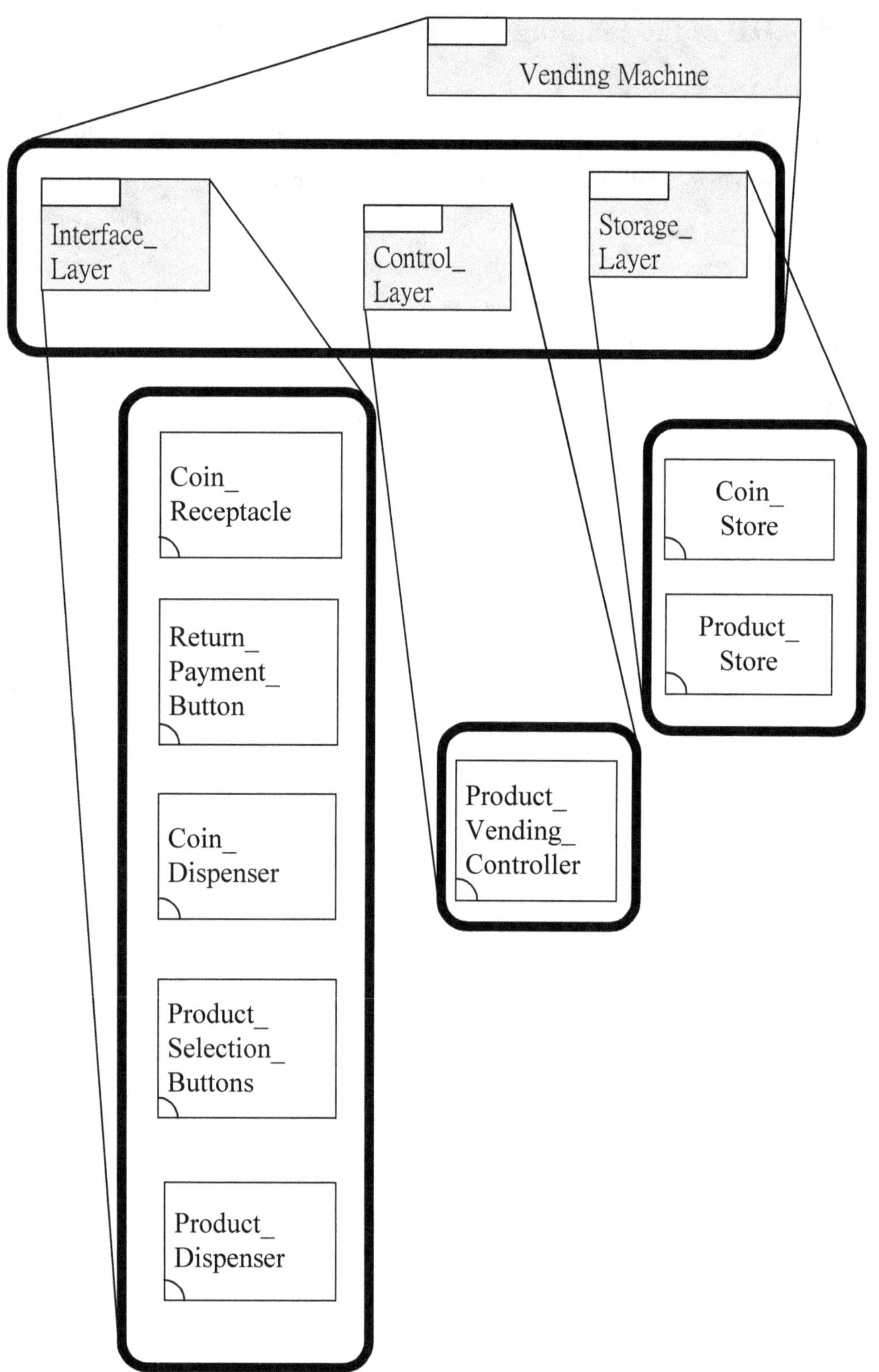

Figure 5-1 AHD of the *Vending Machine*

In Figure 5-1, *Vending Machine, Interface_Layer*, *Control_Layer* and

Storage_Layer are aggregated systems while *Coin_Receptacle*, *Return_Payment_Button*, *Coin_Dispenser*, *Product_Selection_Buttons*, *Product_Dispenser*, *Product_Vending_Controller*, *Coin_Store* and *Product_Store* are non-aggregated systems.

Chapter 6: FD of the Vending Machine

FD is the framework diagram we obtain after the architecture construction is finished. Figure 6-1 shows a FD of the *Vending Machine*.

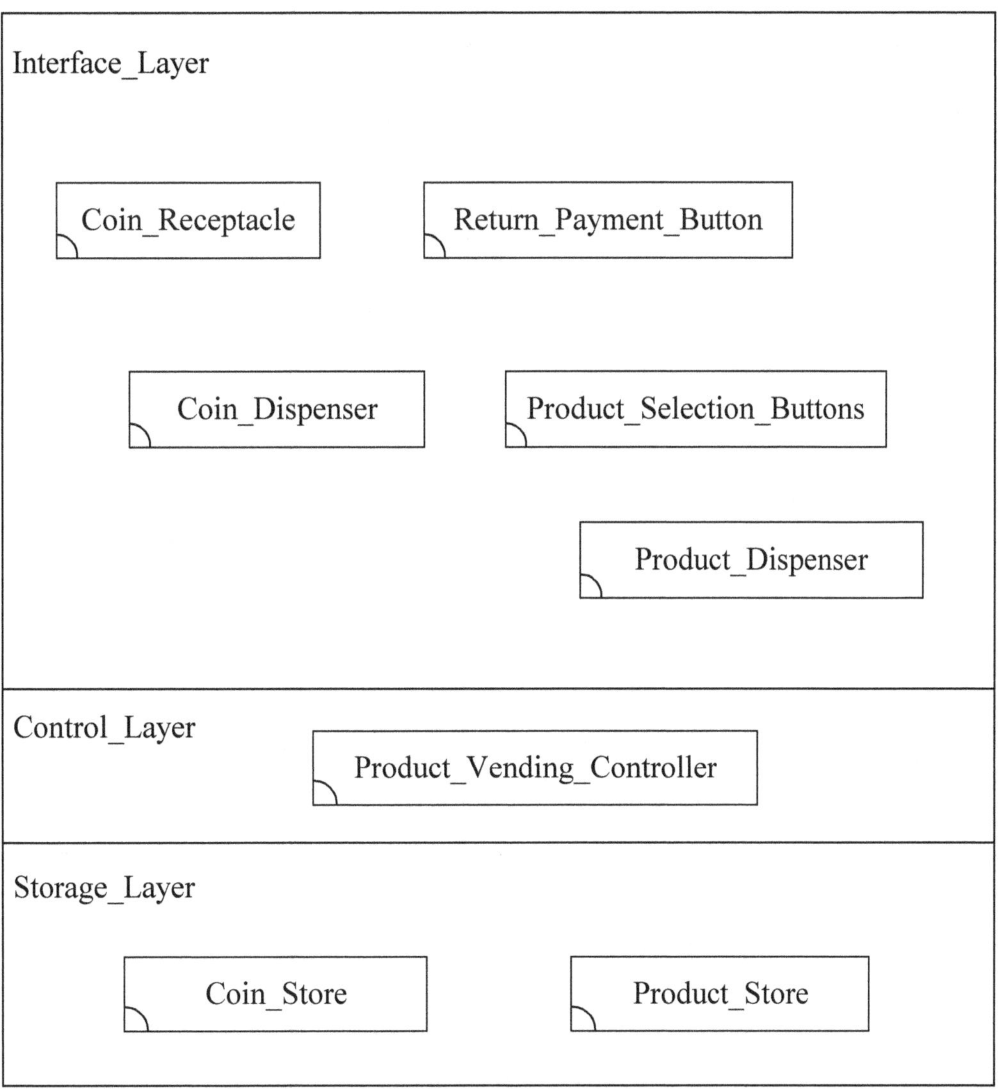

Figure 6-1 FD of the *Vending Machine*

In the above figure, *Interface_Layer* contains the *Coin_Receptacle*, *Return_Payment_Button*, *Coin_Dispenser*, *Product_Selection_Buttons* and *Product_Dispenser* components; *Control_Layer* contains the *Product_Vending_Controller* component; *Storage_Layer* contains the *Coin_Store* and *Product_Store* components.

Chapter 7: CChD of the Vending Machine

CChD is the component channel diagram we obtain after the architecture construction is finished. Figure 7-1 shows a CChD of the *Vending Machine*. In the figure, component *Coin_Receptacle* has one channel: *Accept_Coin*; component *Return_Payment_Button* has one channel: *Return_Payment_Request*; component *Coin_Dispenser* has one channel: *Deliver_Coin*; component *Product_Selection_Buttons* has one channel: *Selection_Request*; component *Product_Dispenser* has one channel: *Deliver_Product*; component *Product_Vending_Controller* has four channels: *Accumulate_Payment*, *Refresh_Selectable_Buttons*, *Return_Payment* and *Product_Select*; component *Coin_Store* has four channels: *Deposit_Coin*, *Return_Coin*, *Dispense_Coin* and *Refill_Change_Coin*; component *Product_Store* has three channels: *Pick_Product*, *Dispense_Product* and *Refill_Vending_Product*.

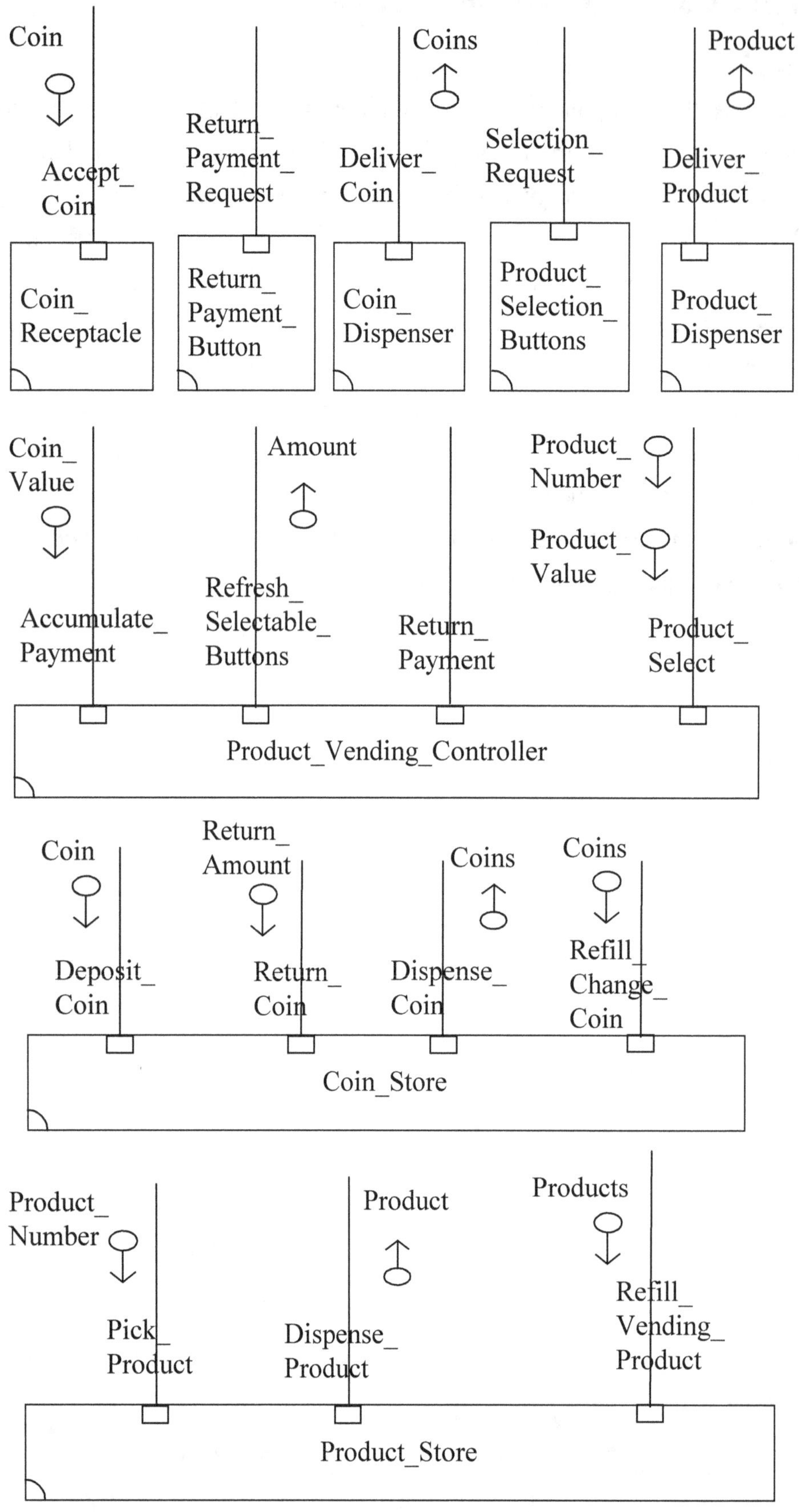

Figure 7-1 CChD of the *Vending Machine*

The channel formula of *Accept_Coin* is *Accept_Coin(In Coin)*. The channel formula of *Return_Payment_Request* is *Return_Payment_Request*. The channel formula of *Deliver_Coin* is *Deliver_Coin(Out Coins)*. The channel formula of *Selection_Request* is *Selection_Request*. The channel formula of *Deliver_Product* is *Deliver_Product(Out Product)*. The channel formula of *Accumulate_Payment* is *Accumulate_Payment(In Coin_Value)*. The channel formula of *Refresh_Selectable_Buttons* is *Refresh_Selectable_Buttons(Out Amount)*. The channel formula of *Return_Payment* is *Return_Payment*. The channel formula of *Product_Select* is *Product_Select(In Product_Number, Product_Value)*. The channel formula of *Deposit_Coin* is *Deposit_Coin(In Coin)*. The channel formula of *Return_Coin* is *Return_Coin(In Return_Amount)*. The channel formula of *Dispense_Coin* is *Dispense_Coin(Out Coins)*. The channel formula of *Refill_Change_Coin* is *Refill_Change_Coin(In Coins)*. The channel formula of *Pick_Product* is *Pick_Product(In Product_Number)*. The channel formula of *Dispense_Product* is *Dispense_Product(Out Product)*. The channel formula of *Refill_Vending_Product* is *Refill_Vending_Product(In Products)*.

Figure 7-2 shows the primitive data type specification of the *Coin*, *Coin_Value*, *Product_Number*, *Product_value*, *Return_Amount* input parameters and *Amount*, *Product* output parameters.

Parameter	Data Type	Instances
Coin	Currency	nickel, dime, quarter
Coin_Value	Integer	5, 10, 25
Product_Number	Integer	21, 22, 23, 24
Product_Value	Integer	50, 60, 70
Return_Amount	Integer	40, 60, 80
Product	Food	coke, milk, water
Amount	Integer	80, 120, 170

Figure 7-2 Primitive Data Type Specification

Figure 7-3 shows the composite data type specification of the *Coins* output parameter occurring in the *Deliver_Coin(Out Coins)*, *Dispense_Coin(Out Coins)* channel formulas and the input parameter occurring in the *Refill_Change_Coin(In Coins)* channel formula.

Parameter	*Coins*
Data Type	ARRAY of Coin: Currency End ARRAY ;
Instances	Coin nickel nickel dime quarter quarter

Figure 7-3 Composite Data Type Specification of *Coins*

Figure 7-4 shows the composite data type specification of the *Products* input parameter occurring in the *Refill_Vending_Product(In Products)* channel formula.

Parameter	*Products*
Data Type	ARRAY of Product: Food End ARRAY ;
Instances	Products coke coke coke milk water

Figure 7-4 Composite Data Type Specification of *Products*

Chapter 8: CCoD of the Vending Machine

CCoD is the component connection diagram we obtain after the architecture construction is finished. Figure 8-1 shows a CCoD of the *Vending Machine*.

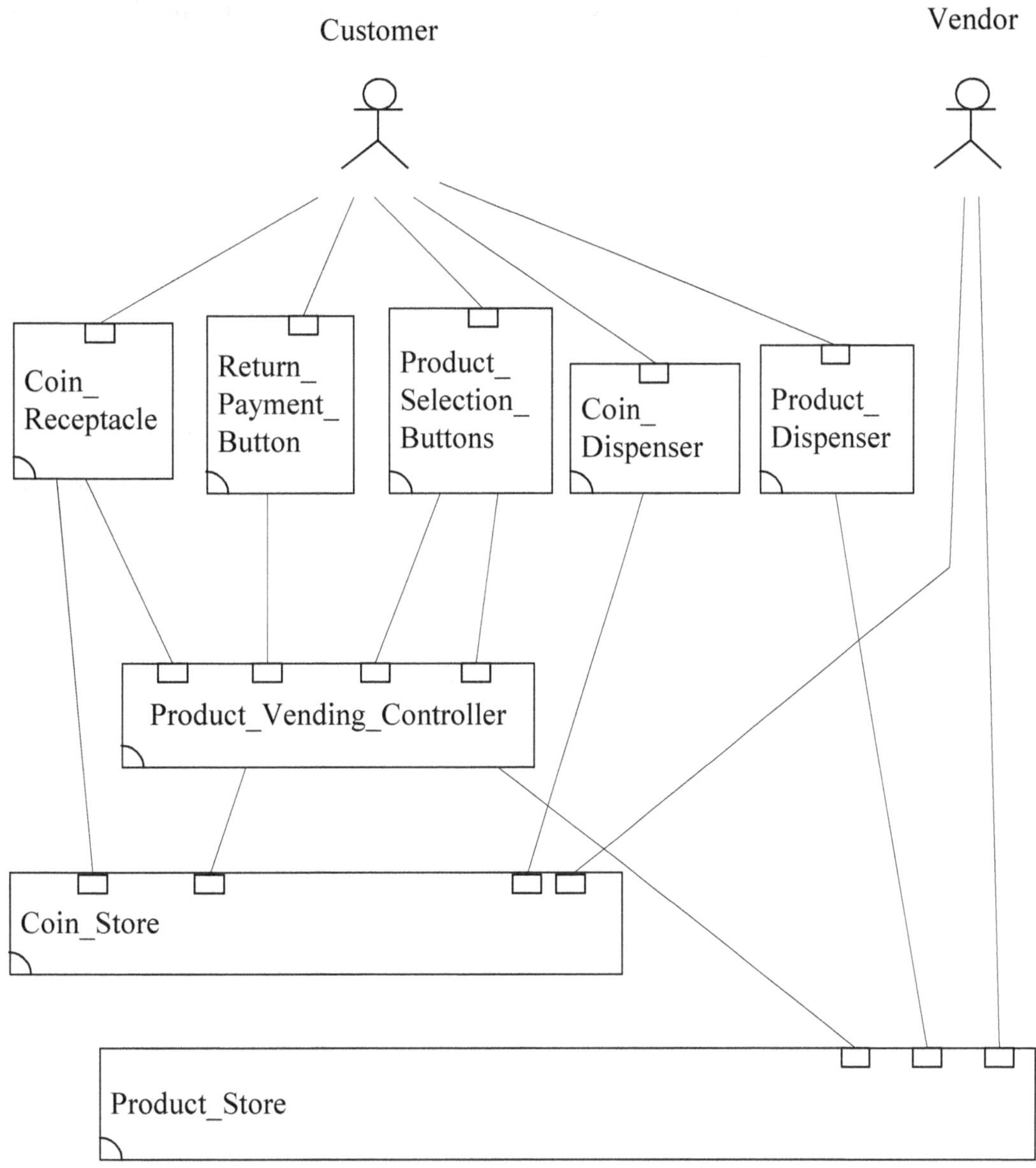

Figure 8-1 CCoD of the *Vending Machine*

In the above figure, actor *Customer* has a connection with each one of the *Coin_Receptacle*, *Return_Payment_Button*, *Product_Selection_Buttons*, *Coin_Dispenser*, *Product_Dispenser* components; actor *Vendor* has a connection with each one of the *Coin_Store*, *Product_Store* components; component *Coin_Receptacle* has a connection with each one of the *Coin_Store*, *Product_Vending_Controller* components; component *Return_Payment_Button* has one connection with the *Product_Vending_Controller* component; component *Product_Selection_Buttons* has two connections with the *Product_Vending_Controller* component; component *Coin_Dispenser* has a connection with the *Coin_Store* component;. component *Product_Dispenser* has a connection with the *Product_Store* component; component *Product_Vending_Controller* has a connection with each one of the *Coin_Store*, *Product_Store* components.

Chapter 9: SBCD of the Vending Machine

SBCD is the structure-behavior coalescence diagram we obtain after the architecture construction is finished. Figure 9-1 shows a SBCD of the *Vending Machine* in which interactions among the *Customer*, *Vendor* actors and the *Coin_Receptacle*, *Return_Payment_Button*, *Coin_Dispenser*, *Product_Selection_Buttons*, *Product_Dispenser*, *Product_Vending_Controller*, *Coin_Store* *Product_Store* components shall draw forth the *Getting_Customer_Payment*, *Returning_Customer_Payment*, *Delivering_Customer_Selection*, *Refilling_Product_Store*, *Refilling_Coin_Store* behaviors.

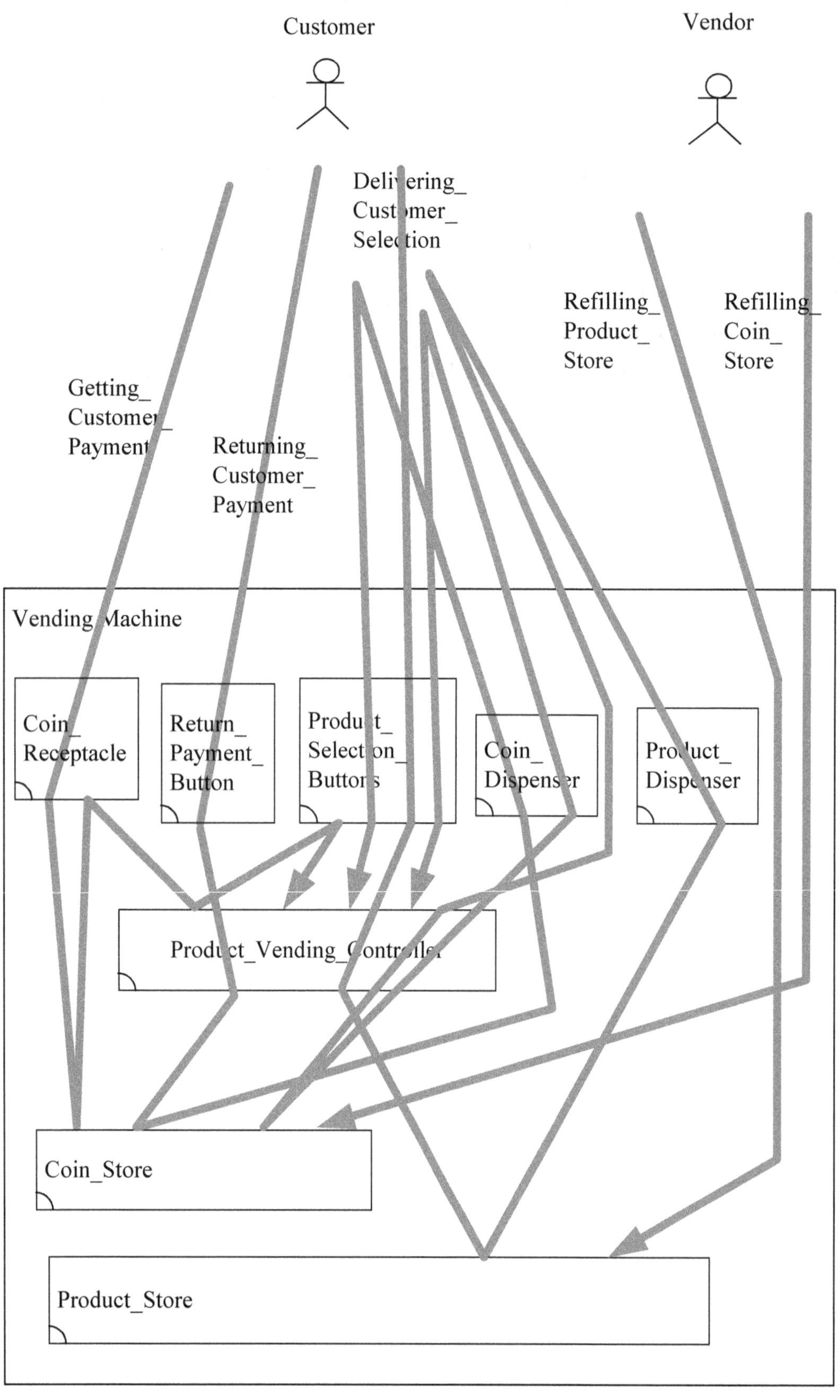

Figure 9-1 SBCD of the *Vending Machine*

The overall behavior of the *Vending Machine* includes the *Getting_Customer_Payment*, *Returning_Customer_Payment*, *Delivering_Customer_Selection*, *Refilling_Product_Store*, *Refilling_Coin_Store* behaviors. In other words, the *Getting_Customer_Payment*, *Returning_Customer_Payment*, *Delivering_Customer_Selection*, *Refilling_Product_Store*, *Refilling_Coin_Store* behaviors together provide the overall behavior of the *Vending Machine*.

The major purpose of adopting the architectural approach, instead of separating the structure model from the behavior model, is to achieve one single coalesced model. In Figure 9-1, hardware architects are able to see that the hardware structure and hardware behavior coexist in the SBCD. That is, in the SBCD of *Vending Machine*, systems architects not only see its hardware structure but also see (at the same time) its hardware behavior.

Chapter 10: IFD of Vending Machines

IFDs are the interaction flow diagrams we obtain after the architecture construction is finished. The overall behavior of *Vending Machine* includes five individual behaviors: *Getting_Customer_Payment*, *Returning_Customer_Payment*, *Delivering_Customer_Selection*, *Refilling_Product_Store*, *Refilling_Coin_Store*. Each individual behavior is represented by an execution path. We use an IFD to define each one of these execution paths.

Figure 10-1 shows an IFD of the *Getting_Customer_Payment* behavior. First, actor *Customer* interacts with the *Coin_Receptacle* component through the *Accept_Coin* channel interaction, carrying the *Coin* input parameter. Next, component *Coin_Receptacle* interacts with the *Coin_Store* component through the *Deposit_Coin* channel interaction, carrying the *Coin* input parameter. Continuingly, component *Coin_Receptacle* interacts with the *Product_Vending_Controller* component through the *Accumulate_Payment* channel interaction, carrying the *Coin_Value* input parameter. Finally, component *Product_Selection_Buttons* interacts with the *Product_Vending_Controller* component through the *Refresh_Selectable_Buttons* channel interaction, carrying the *Amount* output parameter.

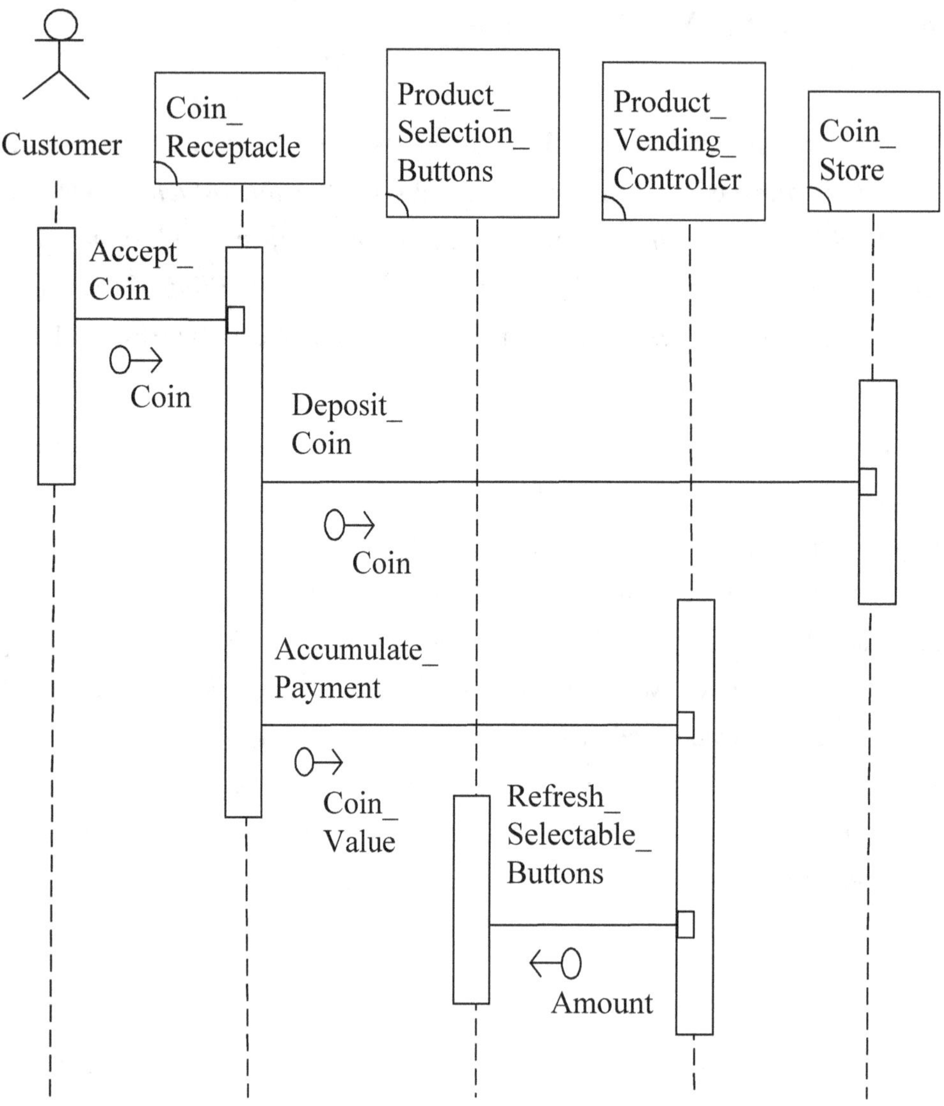

Figure 10-1 IFD of the *Getting_Customer_Payment* Behavior

Figure 10-2 shows an IFD of the *Returning_Customer_Payment* behavior. First, actor *Customer* interacts with the *Return_Payment_Button* component through the *Return_Payment_Request* channel interaction. Next, component *Return_Payment_Button* interacts with the *Product_Vending_Controller* component through the *Return_Payment* channel interaction. Continuingly, component *Product_Vending_Controller* interacts with the *Coin_Store* component through the *Return_Coin* channel interaction, carrying the *Return_Amount* input parameter. Continuingly, component *Coin_Dispenser* interacts with the *Coin_Store* component through the *Dispense_Coin* channel interaction, carrying the *Coins* output parameter. Continuingly, actor *Customer* interacts with the *Coin_Dispenser* component through the *Deliver_Coin* channel interaction, carrying the *Coins* output parameter. Finally, component *Product_Selection_Buttons* interacts with the *Product_Vending_Controller* component through the *Refresh_Selectable_Buttons* channel interaction, carrying the *Amount* output parameter.

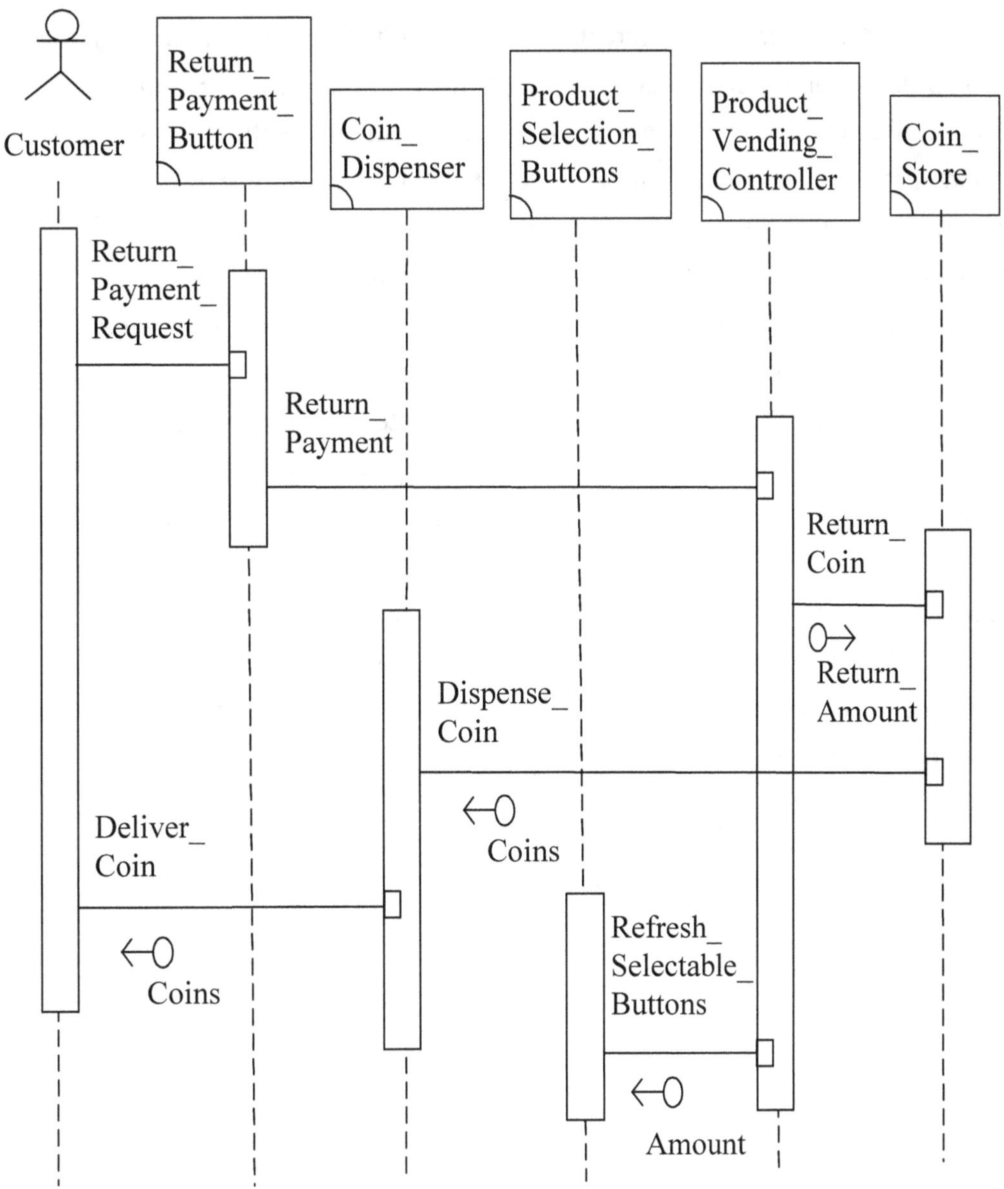

Figure 10-2 IFD of the *Returning_Customer_Payment* Behavior

Figure 10-3 shows an IFD of the *Delivering_Customer_Selection* behavior. First, actor *Customer* interacts with the *Product_Selection_Buttons* component through the *Selection_Request* channel interaction. Next, component *Product_Selection_Buttons* interacts with the *Product_Vending_Controller* component through the *Product_Select* channel interaction, carrying the *Product_Number*, *Product_Value* input parameters. Continuingly, component *Product_Vending_Controller* interacts with the *Product_Store* component through the *Pick_Product* channel interaction, carrying the *Product_Number* input parameter. Continuingly, component *Product_Dispenser* interacts with the *Product_Store* component through the *Dispense_Product* channel interaction, carrying the *Product* output parameter. Next, actor *Customer* interacts with the *Product_Dispenser* component through the *Deliver_Product* channel interaction, carrying the *Product* output parameters. Continuingly, component *Product_Vending_Controller* interacts with the *Coin_Store* component through the *Return_Coin* channel interaction, carrying the *Return_Amount* input parameter. Continuingly, component *Coin_Dispenser* interacts with the *Coin_Store* component through the *Dispense_Coin* channel interaction, carrying the *Coins* output parameter. Next, actor *Customer* interacts with the *Coin_Dispenser* component through the *Deliver_Coin* channel interaction, carrying the *Coins* output parameters. Finally, component *Product_Selection_Buttons* interacts with the *Product_Vending_Controller* component through the *Refresh_Selectable_Buttons* channel interaction, carrying the *Amount* output parameter.

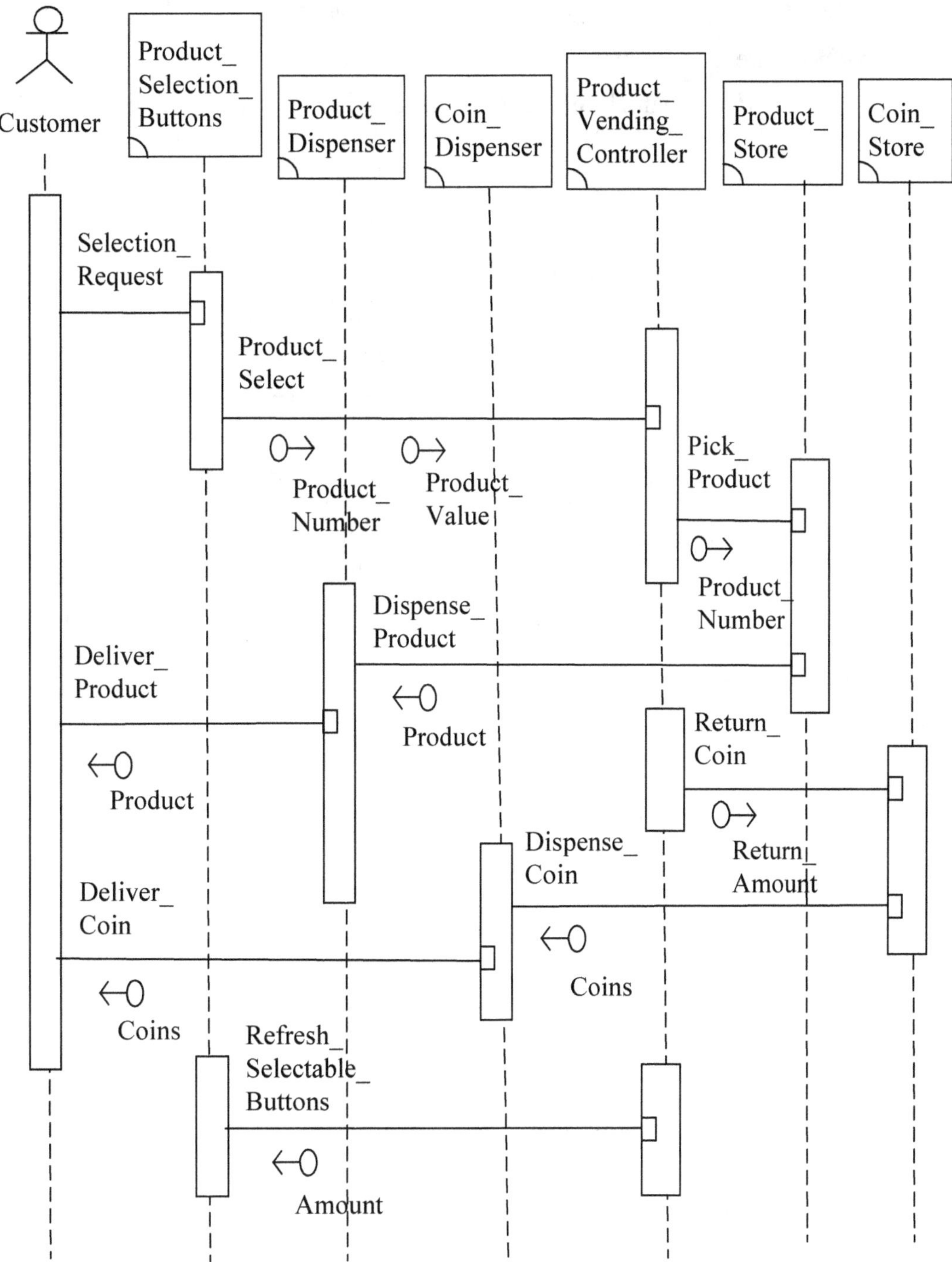

Figure 10-3 IFD of the *Delivering_Customer_Selection* Behavior

Figure 10-4 shows an IFD of the *Refilling_Product_Store* behavior. First, actor *Vendor* interacts with the *Product_Store* component through the *Refill_Vending_Product* channel interaction, carrying the *Products* input parameter.

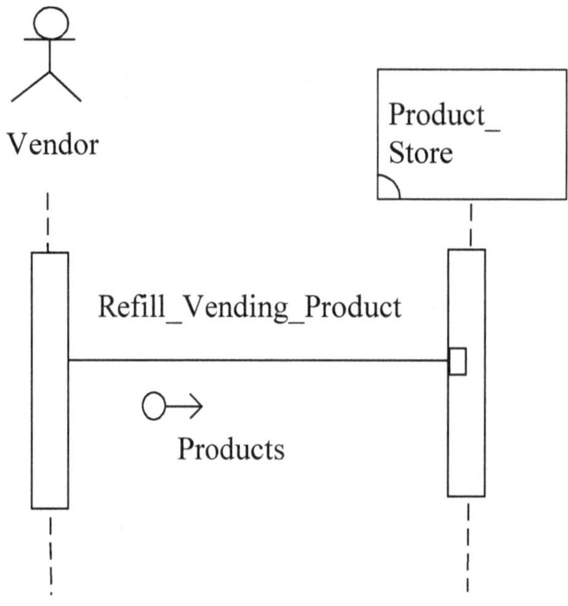

Figure 10-4 IFD of the *Refilling_Product_Store* Behavior

Figure 10-5 shows an IFD of the *Refilling_Coin_Store* behavior. First, actor *Vendor* interacts with the *Coin_Store* component through the *Refill_Change_Coin* channel interaction, carrying the *Coins* input parameter.

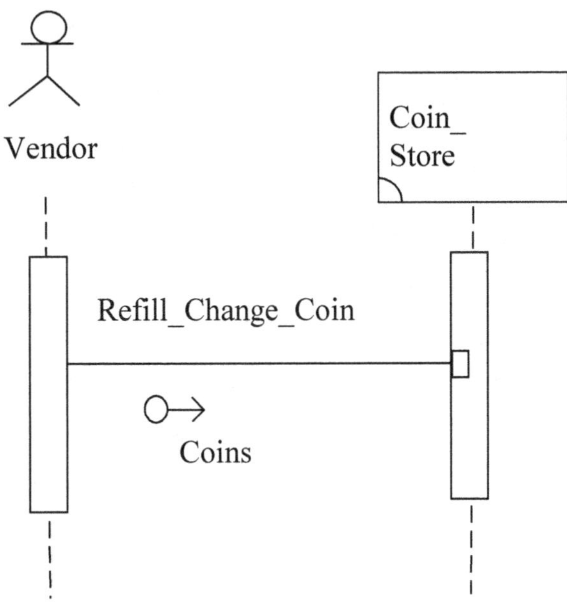

Figure 10-5 IFD of the *Refilling_Coin_Store* Behavior

APPENDIX A: SBC ARCHITECTURE DESCRIPTION LANGUAGE

(1) Architecture Hierarchy Diagram

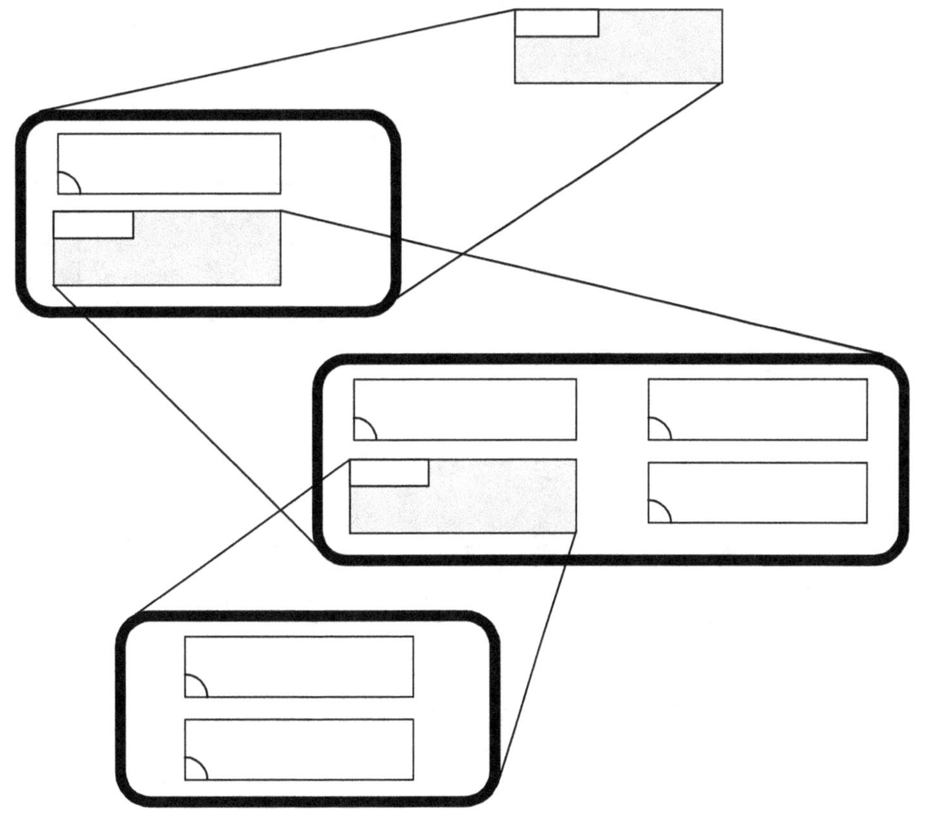

: Aggregated System

: Non-Aggregated System, Component

(2) Framework Diagram

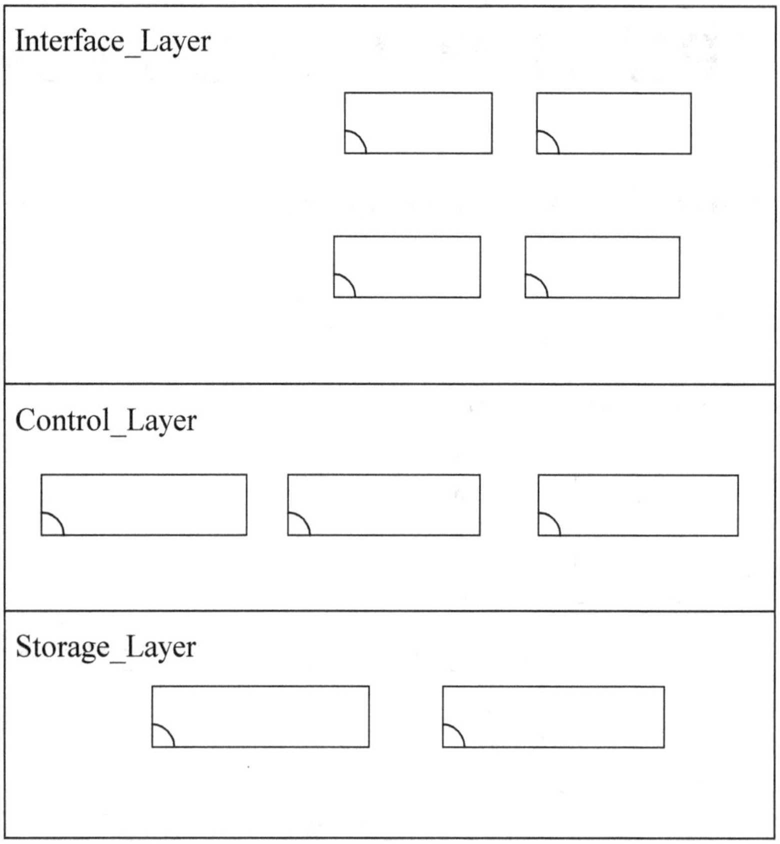

Interface_Layer

Control_Layer

Storage_Layer

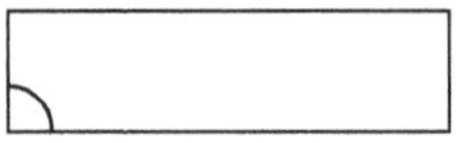

: Component

(3) Component Channel Diagram

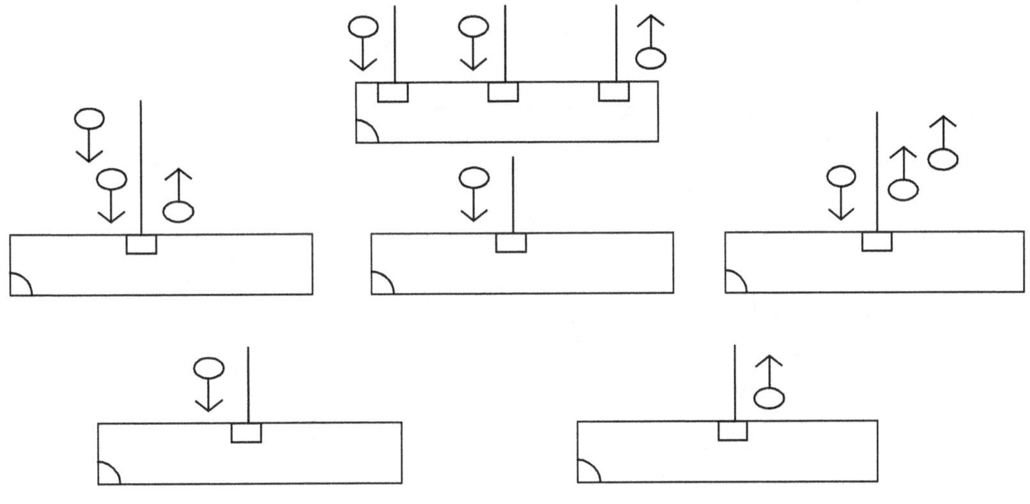

: Channel

: Input Data

: Output Data

: Component

(4) Component Connection Diagram

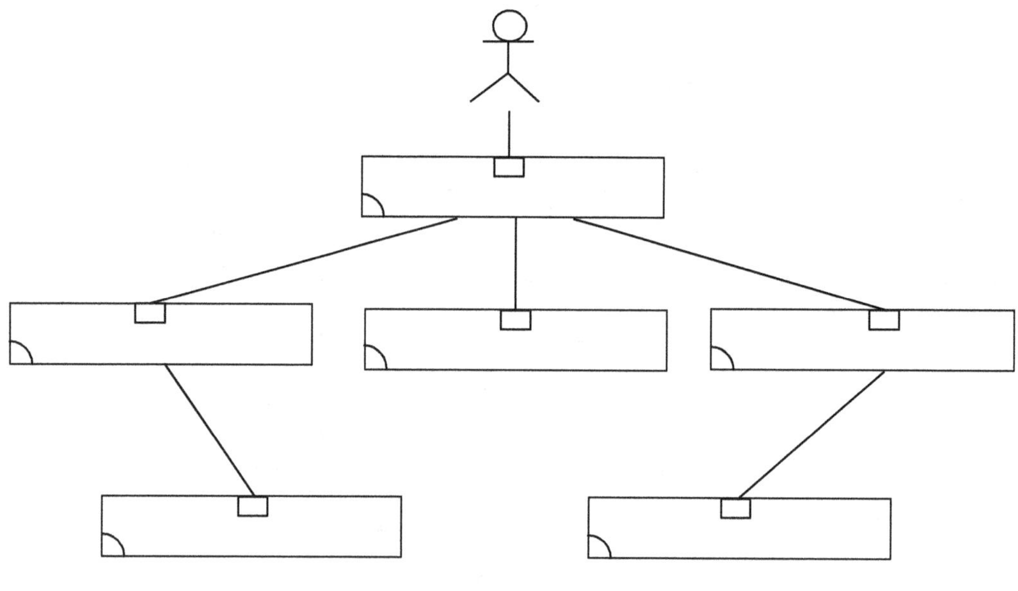

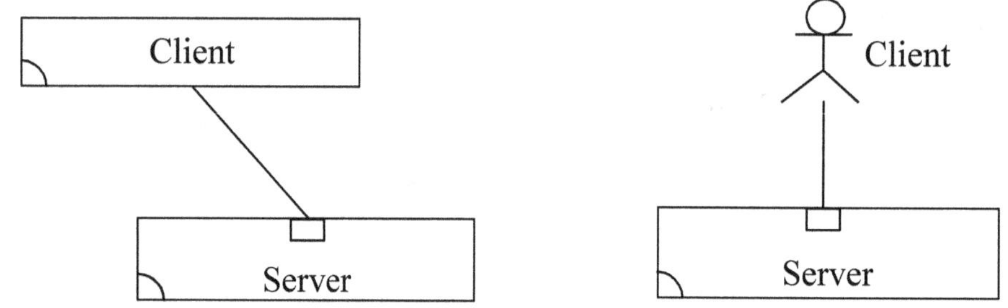

(5) Structure-Behavior Coalescence Diagram

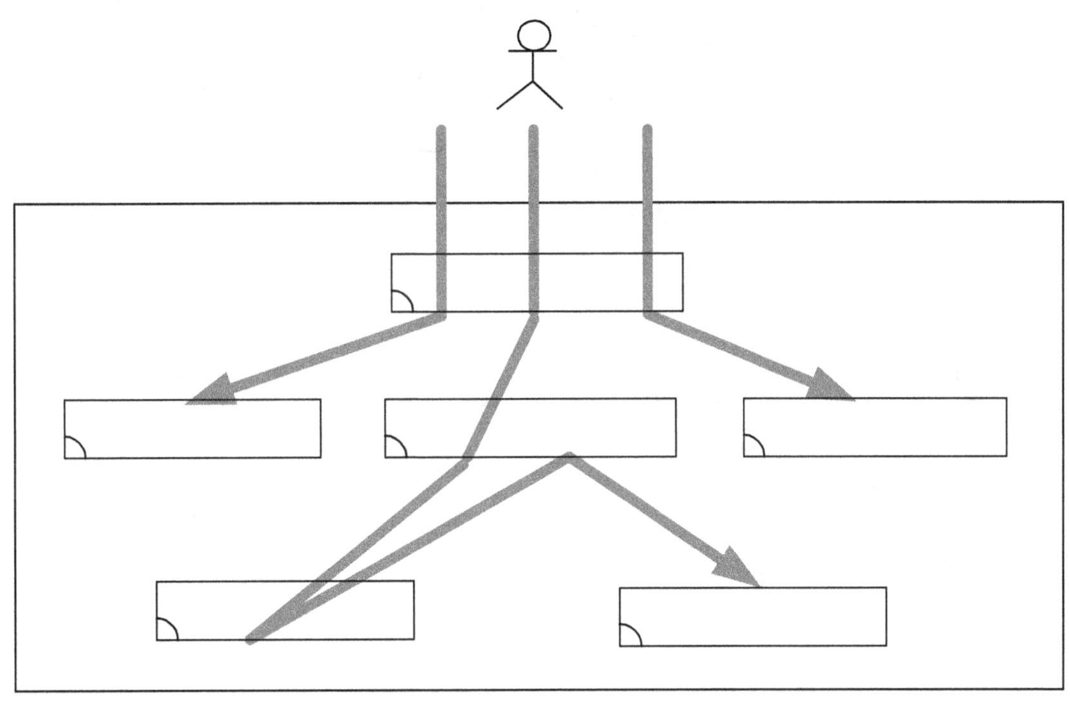

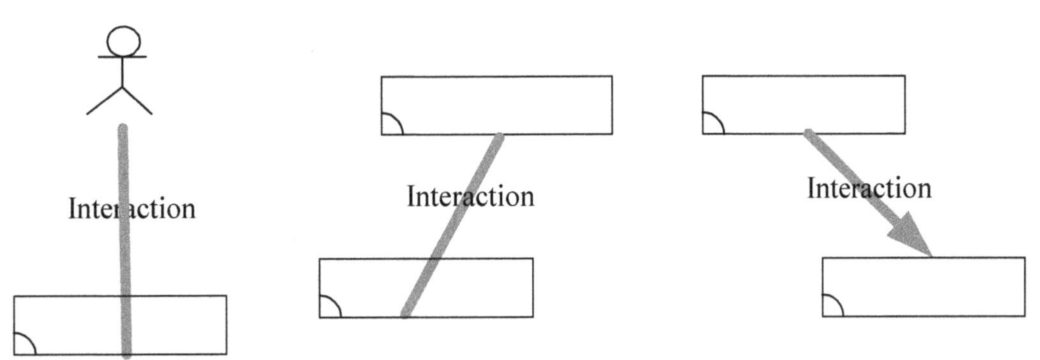

(6) Interaction Flow Diagram

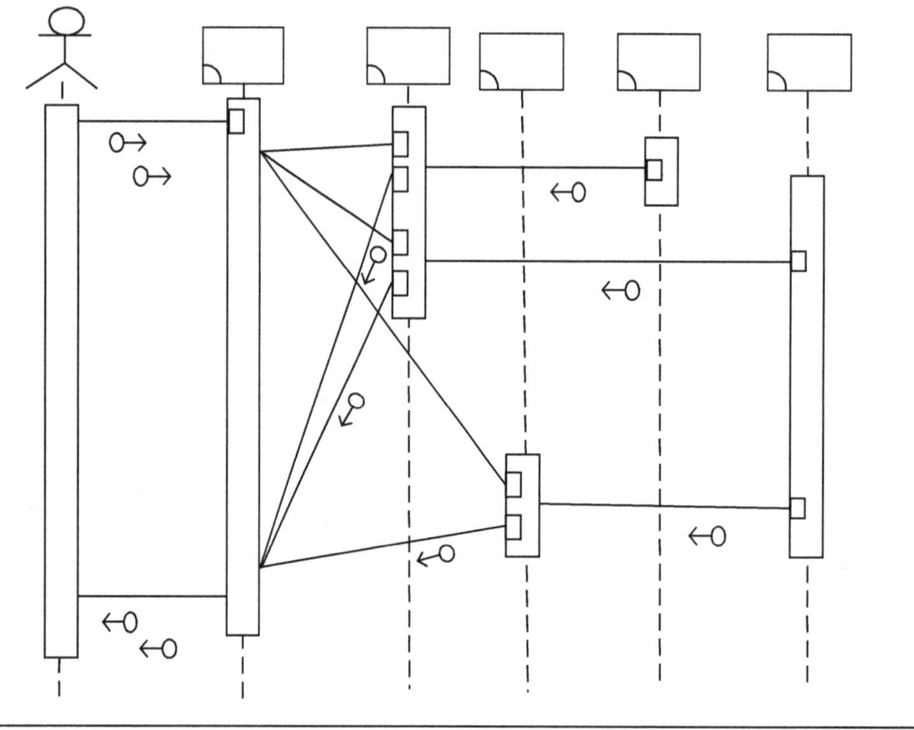

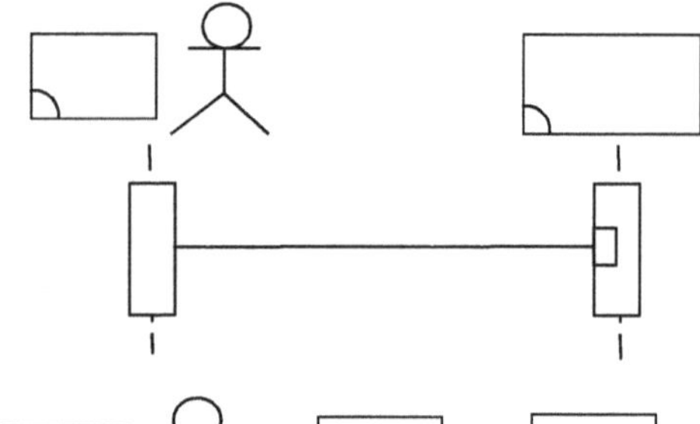

: Channel Interaction

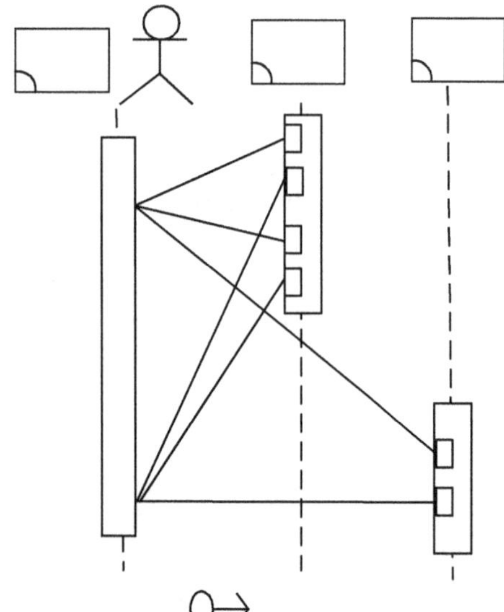

: Conditional
Channel Interaction

: Input Data

: Output Data

APPENDIX B: SBC PROCESS ALGEBRA

(1) Channel-Based Single-Queue SBC Process Algebra

(1) <System> ::= "**fix**(" <Process_Variable> "="<IFD> "• " <Process_Variable>
{"+" <IFD> " • " <Process_Variable>} ")"

(2) <IFD> ::= <Type_1_Interaction> {"• " <Type_1_Or_2_Interaction>}

(3) <Type_1_Or_2_Interaction> ::= <Type_1_Interaction>

| <Type_2_Interaction>

(2) Channel-Based Multi-Queue SBC Process Algebra

(1) <System> ::= <FixIFD> {"|| " <FixIFD>}

(2) <FixIFD> ::= "**fix**(" <Process_Variable>"="<IFD>
"•" <Process_Variable> ")"

(3) <IFD> ::= <Type_1_Interaction> {"• " Type_1_Or_2_Interaction>}

(4) <Type_1_Or_2_Interaction> ::= <Type_1_Interaction>

| <Type_2_Interaction>

(3) Channel-Based Infinite-Queue SBC Process Algebra

(1) <System> ::= "**!** ("<IFD> " • " *STOP* ")" {"|| **!** (" <IFD> " • " *STOP* ")"}

(2) <IFD> ::= <Type_1_Interaction> {"• " <Type_1_Or_2_Interaction>}

(3) <Type_1_Or_2_Interaction> ::= <Type_1_Interaction>

| <Type_2_Interaction>

BIBLIOGRAPHY

[Bern09] Bernstein, D. et al., "Blueprint for the Intercloud – Protocols and Formats for Cloud Computing Interoperability," *IEEE Computer Society*, 2009, pp.328-336.

[Burd10] Burd, S. D., *Systems Architecture*, 6th Edition, Cengage Learning, 2010.

[Chao14a] Chao, W. S., *Systems Thingking 2.0: Architectural Thinking Using the SBC Architecture Description Language*, CreateSpace Independent Publishing Platform, 2014.

[Chao14b] Chao, W. S., *General Systems Theory 2.0: General Architectural Theory Using the SBC Architecture*, CreateSpace Independent Publishing Platform, 2014.

[Chao14c] Chao, W. S., *Systems Modeling and Architecting: Structure-Behavior Coalescence for Systems Architecture*, CreateSpace Independent Publishing Platform, 2014.

[Chao15a] Chao, W. S., *A Process Algebra For Systems Architecture: The Structure-Behavior Coalescence Approach*, CreateSpace Independent Publishing Platform, 2015.

[Chao15b] Chao, W. S., *An Observation Congruence Model For Systems Architecture: The Structure-Behavior Coalescence Approach*, CreateSpace Independent Publishing Platform, 2015.

[Chao16] Chao, W. S., *System: Contemporary Concept, Definition, and Language*, CreateSpace Independent Publishing Platform, 2016.

[Chao17a] Chao, W. S., *Channel-Based Single-Queue SBC Process Algebra For Systems Definition: General Architectural Theory at Work*, CreateSpace

Independent Publishing Platform, 2017.

[Chao17b] Chao, W. S., *Channel-Based Multi-Queue SBC Process Algebra For Systems Definition: General Architectural Theory at Work*, CreateSpace Independent Publishing Platform, 2017.

[Chao17c] Chao, W. S., *Channel-Based Infinite-Queue SBC Process Algebra For Systems Definition: General Architectural Theory at Work*, CreateSpace Independent Publishing Platform, 2017.

[Chao17d] Chao, W. S., *Operation-Based Single-Queue SBC Process Algebra For Systems Definition: General Architectural Theory at Work*, CreateSpace Independent Publishing Platform, 2017.

[Chao17e] Chao, W. S., *Operation-Based Multi-Queue SBC Process Algebra For Systems Definition: Unification of Systems Structure and Systems Behavior*, CreateSpace Independent Publishing Platform, 2017.

[Chao17f] Chao, W. S., *Operation-Based Infinite-Queue SBC Process Algebra For Systems Definition: Unification of Systems Structure and Systems Behavior*, CreateSpace Independent Publishing Platform, 2017.

[Chec99] Checkland, P., *Systems Thinking, Systems Practice: Includes a 30-Year Retrospective*, 1st Edition, Wiley, 1999.

[Craw15] Crawley, P. et al., *System Architecture: Strategy and Product Development for Complex Systems*, Prentice Hall, 2015.

[Dam06] Dam, S., *DoD Architecture Framework: A Guide to Applying System Engineering to Develop Integrated Executable Architectures*, BookSurge Publishing, 2006.

[Date03] Date, C. J., *An Introduction to Database Systems*, 8th Edition, Addison Wesley, 2003.

[Denn08] Dennis, A. et al., *Systems Analysis and Design*, 4th Edition, Wiley, 2008.

[Dori95] Dori, D., "Object-Process Analysis: Maintaining the Balance between System Structure and Behavior," *Journal of Logic and Computation* 5(2), pp.227-249, 1995.

[Dori02] Dori, D., *Object-Process Methodology: A Holistic Systems Paradigm*, Springer Verlag, New York, 2002.

[Dori16] Dori, D., *Model-Based Systems Engineering with OPM and SysML*, Springer Verlag, New York, 2016.

[Elma10] Elmasri, R., *Fundamentals of Database Systems*, 6th Edition, Addison Wesley, 2010.

[Hoar85] Hoare, C. A. R., *Communicating Sequential Processes*, Prentice-Hall, 1985.

[Kend10] Kendall, K. et al., *Systems Analysis and Design*, 8th Edition, Prentice Hall, 2010.

[Maie09] Maier, M. W., *The Art of Systems Architecting*, 3rd Edition, CRC Press, 2009.

[Miln89] Milner, R., *Communication and Concurrency*, Prentice-Hall, 1989.

[Miln99] Milner, R., *Communicating and Mobile Systems: the π-Calculus*, 1st Edition, Cambridge University Press, 1999.

[O'Rou03] O'Rourke, C. et al, *Enterprise Architecture Using the Zachman Framework*, 1st Edition, Course Technology, 2003.

[Pele00] Peleg, M. et al., "The Model Multiplicity Problem: Experimenting with Real-Time Specification Methods". *IEEE Tran. on Software Engineering*. 26 (8), pp. 742–759, 2000.

[Prat00] Pratt, T. W. et al., *Programming Languages: Design and Implementation*, 4th Edition, Prentice Hall 2000.

[Pres09] Pressman, R. S., *Software Engineering: A Practitioner's Approach*, 7th Edition, McGraw-Hill, 2009.

[Putm00] Putman, J. R. et al., *Architecting with RM-ODP*, Prentice-Hall, 2000.

[Rayn09] Raynard, B., *TOGAF The Open Group Architecture Framework 100 Success Secrets*, Emereo Pty Ltd, 2009.

[Roza11] Rozanski, N. et al., *Software Systems Architecture: Working With Stakeholders Using Viewpoints and Perspectives*, 2nd Edition, Addison-Wesley Professional, 2011.

[Rumb91] Rumbaugh, J. et al., *Object-Oriented Modeling and Design*, Prentice-Hall, 1991.

[Segr02] Segrave, K., *Vending Machines: An American Social History*, McFarland & Company, 2002.

[Seth96] Sethi, R., *Programming Languages: Concepts and Constructs*, 2nd Edition, Addison-Wesley, 1996.

[Sode03] Soderborg, N.R. et al., "OPM-based Definitions and Operational Templates," *Communications of the ACM* 46(10), pp. 67-72, 2003.

[Somm06] Sommerville, I., *Software Engineering*, 8th Edition, Addison-Wesley, 2006.

[Toga08] The Open Group, *TOGAF Version 9 - A Manual (TOGAF Series)*, 9th Edition, Van Haren Publishing, 2008.

[Your99] Yourdon, E., *Death March: The Complete Software Developer's Guide to Surviving Mission Impossible Projects*, Prentice-Hall, 1999.

www.ingramcontent.com/pod-product-compliance
Lightning Source LLC
Chambersburg PA
CBHW060009210526
45170CB00017B/2116

* 9 7 8 1 5 4 6 7 5 3 0 5 6 *